ICIAM2023 Springer Series

Volume 4

This series aims to publish some of the most relevant results presented at the ICIAM 2023 held in Tokyo in August 2023.

The series is managed by an independent Editorial Board, and will include peer-reviewed content only, including the Invited Speakers volume as well as books resulting from mini-symposia and collateral workshops.

The series is aimed at providing useful reference material to academic and researchers at an international level.

Mei Kobayashi

Federated Learning

A Primer for Mathematicians

Springer

Mei Kobayashi
EAGLYS
Shibuya, Tokyo, Japan

ISSN 3091-3098 ISSN 3091-3101 (electronic)
ICIAM2023 Springer Series
ISBN 978-981-96-9222-4 ISBN 978-981-96-9223-1 (eBook)
https://doi.org/10.1007/978-981-96-9223-1

Mathematics Subject Classification: 68-06, 68W99, 68P25, 68-01, 68P27, 68T07

This Springer imprint is published by the registered company Springer Nature Singapore Pte Ltd.
The registered company address is: 152 Beach Road, #21-01/04 Gateway East, Singapore 189721, Singapore

To my Mother for her love and support.
Happy 100th Birthday ... and many more!

Foreword

This monograph covers a wide range of concepts from mathematics to systems. The content is written using well-defined technical terms of the artificial intelligence (AI) area and is easily understandable. This monograph is particularly recommended for mathematicians and scientists who are interested in collaborating with software engineers and project managers to develop AI systems. It is a valuable tool for getting up to speed and acquiring basic knowledge of common technical terms and the minimum mathematical background in the AI area to enable fruitful technical communication with colleagues in multidisciplinary projects.

I hope readers will enjoy this friendly little book and create useful AI systems to improve society and our quality of life.

Tokyo, Japan
October 2024

Yoshihiro Ohba
Chief Specialist, KIOXIA Corporation

Preface

The idea for this tutorial on federated learning arose after a mini-symposium at the International Congress on Industrial and Applied Mathematics (ICIAM) 2023 in Tokyo. The session, *Secure Computing: Maintaining Personal Privacy while Analyzing Data*,[1] featured distinguished researchers in engineering and computer science. Professor Hayato Yamana of Waseda University, Dr. Yukio Uematsu from Tokyo University of Science and Nokia, and Dr. Tsubasa Takahashi from LINE Yahoo Corporation kindly accepted my invitation to participate in a conference consisting primarily of applied mathematicians. Their spirit of interdisciplinary collaboration came through in their friendly and open discussions during the Q&A session following their talks.

Computer scientists, including the session speakers, have been developing reliable and efficient methods for encrypting, transmitting, and analyzing sensitive data while protecting the personal privacy of individuals and proprietary information of institutions. This rapidly growing field, known as *secure computing*, benefits from collaborations between mathematicians, computer scientists, owners of massive data sets, and prospective users of secure computing systems.

This monograph is a primer on a secure computing framework, known as *federated learning*. Federated learning is the study of methods to enable multiple parties to collaboratively train machine learning artificial intelligence (AI) models, while each party retains its own, raw data on-premise, never sharing it with others.

Many of the ideas in federated learning were originally proposed, designed, and implemented independently by Chilimbi et al. as "*a large-scale commodity distributed system* (that) *can be used to efficiently train very large distributed neural networks (DNNs)*"[2] and Shokri et al. as "*a practical system that enables multiple*

[1] ICIAM 2023 Mini-Symposium, *Secure Computing: Maintaining Personal Privacy while Analyzing Data*: iciam2023.org/registered_data?id=00819.

[2] Chilimbi, T., Suzue, Y., Apacible, J., Kalyanaraman, K.: Project Adam (2014) Building an efficient and scalable deep learning training system. USENIX Symposium on Operating Systems Design and Implementation.

parties to jointly learn an accurate neural network model for a given objective without sharing their input datasets".[3]

Federated learning was propelled to fame in the AI research community by scientists at Google who coined the term in a paper[4] describing the development and deployment of a massively scaled-up version on mobile and smart home devices of millions of their customers. The scientists sought a means to understand preferences of individual customers using summaries of their behaviors (not raw user data) that were sent to Google data centers. Google benefitted by: having users pay for computations on edge devices for behavioral analysis, receiving less data volume, and avoiding privacy litigation by receiving results from analysis and not raw data. The federated learning framework has since been extended for use in a wide range of applications and is recognized as a field within the AI domain. So willingly or not, our online behaviors have become subjects of study by companies and service providers on the internet due to federated learning.

This monograph is intended to be accessible to anyone with a background in undergraduate applied mathematics. It covers the basics of topics from computer science that are needed to understand examples of simple federated computing frameworks. It is my hope that by learning basic concepts and technical jargon from computer science, readers will be able to start collaborative work with researchers interested in secure computing.

Chapter 1 gives the background and motivation for data security, federated learning and the simplest type of neural network. The goal is to familiarize readers with definitions and some AI jargon. Chapter 2 introduces the idea of multiparty computation and why enhancements are needed to provide security and privacy. Chapter 3 discusses edge computing, something that cell phones of many readers are doing with or without their knowledge. The material from the first three chapters is useful for understanding Chap. 4, which presents the basics of federated learning. Chapter 5 discusses two examples of attacks on federated learning systems: gradient inversion and data poisoning. The examples show the difficulty in deploying new information sharing frameworks in the real world. Since it is beyond the scope of this work to discuss the myriad of possible cryptographic, security, and privacy attacks, we refer readers to some excellent survey works in the area.

In closing, I would like to thank the organizers of ICIAM 2023; the session speakers; the publication team at Springer; the reviewers of the manuscript; and my colleagues at EAGLYS for their support. In particular, I would like to thank William Hsiao for his encouragement and detailed suggestions for improvement.

Tokyo, Japan
September 2024

Mei Kobayashi

[3] Shokri, R., Shmatikov, V. (2015) Privacy-preserving deep learning. Proceedings of ACM SIGSAC Conference on Computer and Communications Security, 1310–1321.

[4] McMahan, B., Moore, E., Ramage, D., Hampson, S., and B.A. y Arcas (2017) Communication efficient learning of deep networks from decentralized data. Proceedings of International Conference on Artificial Intelligence and Statistics.

Competing Interests The author has no competing interests to declare that are relevant to the content of this manuscript.

Contents

Chapter 1
Introduction

Abstract The volume of digital data has been growing exponentially during the past decade. Institutional decision-makers are interested in recouping associated costs by uncovering actionable insights through traditional data analytics and predictive analysis using artificial intelligence (AI). However, a major impediment to deploying AI is the lack of high-quality training data for producing accurate and robust models of practical value. This chapter begins with a review of statistics and predictions on data from reputable consulting firms. It is followed by an introduction to neurons, the basic building blocks of AI models. The introduction is not intended to be a comprehensive tutorial; only terms and concepts that appear in discussions on machine learning and AI in later chapters will be introduced. This chapter concludes with a quick review on methods for data augmentation and synthetic generation of data. These methods may be useful in some scenarios where data sets need to be enlarged. However, they may not always produce useful data for training AI models. In these cases, federated learning may be a possible solution.

1.1 Background and Motivation

The global volume of digital data has been growing exponentially, doubling every 2 to 3 years [81]. Many consulting firms project that the volume will reach 170–180 zetabytes[1] by 2025, a staggering size which Deloitte describes as: "*a stack of blu-ray disks that could reach the moon 23 times!*".[2]

Institutional decision-makers are interested in recouping costs associated with the generation, storage, and security of their data through the use of data analytics to discover *actionable insights*, i.e., insights that can be connected with actions that

[1] A zetabyte (ZB) is a unit of measurement for computers of the future. One zetabyte is a trillion gigabytes, or a billion trillion (1,000,000,000,000,000,000,000) bytes. 1000 zetabytes make 1 yottabyte.

[2] Deloitte website, "Data: a small four-letter word which has grown exponentially to such a big value". Available online: www2.deloitte.com/cy/en/pages/technology/articles/data-grown-big-value.html. Accessed March 3, 2024.

M. Kobayashi, *Federated Learning*, ICIAM2023 Springer Series 4,
https://doi.org/10.1007/978-981-96-9223-1_1

work to achieve business targets, such as cost reduction, increased productivity, and higher sales volume. Examples include: identifying bottlenecks that increase costs, understanding customer profiles and preferences, identifying sources of customer discontent and customer churn, analyzing the effectiveness of marketing campaigns, and discovering fraud in near real time. According to a report by Fortune Business Insights [31]:

> The global big data analytics market size was valued at USD 307.51 billion in 2023. The market is projected to grow from USD 348.21 billion in 2024 to USD 924.39 billion by 2032, exhibiting a CAGR (compound annual growth rate) of 13.0% during the forecast period.

Although numerical figures may differ, reputable research and consulting firms agree on the overall trend; the *compound annual growth rate* (CAGR) for the global data analytics market size is expected to be well above 10%, with some predicting as high as 30% over the next several years.

Over a decade ago, decision-makers relied on analysis of data from past events to understand their customers, suppliers, business partners, etc. Predictions about the future were based on extrapolation. Recently, decision-makers have been interested in predictive analysis using artificial intelligence (AI) models since they can take into account many parameters simultaneously. However, in many instances, a major impediment is the lack of high-quality training data required to produce AI models with sufficiently high accuracy and robustness to be of practical value. In some cases, the data does not exist (e.g., financial fraud data, medical data for diagnosis of rare diseases), while in others, the expense and time constraints render it extremely difficult or impossible (e.g., polling data for predicting election outcomes and seasonal data).

The purpose of this monograph is to introduce *federated learning*, a framework that enables multiple parties to securely share local AI models—not the raw data, itself—to build a high-quality global AI model. Necessary terminology and concepts will be introduced along the way. In this monograph the term "AI" will be used to denote predictive AI models for business decision-making (e.g., binary classification, categorization, regression) and NOT generative AI methods (such as large-language models for chatbots or image and video generation models) that have been making headline news as of late.

In many publications, the terms *neural networks* and *neural network models* are used synonymously with *AI models*. Generally, the term *AI model* tends to refer to a highly complex, sophisticated neural network, while the term *neural network* can be applied to anything from a small network consisting of one fundamental unit (or *neuron*) to a highly complex network. For the sake of consistency, this monograph will use the term *AI models*, whenever possible. However, in the next two subsections, we will be using terminology associated with neural networks since the discussion will focus on the most fundamental unit (or building block) of neural networks and very small, toy-like examples composed of several units.

This monograph assumes that readers have an undergraduate mathematics background, i.e., basic knowledge of calculus, basic linear algebra, and elementary statistical concepts. Readers who seek to refresh mathematical concepts may find the following helpful: Axler's book [5], Chapter 22 of Werness and Wu's [89] online book, and selected chapters from the website for the fifth edition of Strang's book [79] (all of which are available via free download). Excellent texts for purchase are books by Chollet [17], Deisenrioth [24], Kneusel [50], Strang [79], and Watkins [87].

For readers who seek to review the basics of neural networks and AI models, there are numerous friendly texts such as Glassner [34], Raff [71], and Trask [82]. A bit old, but standard reference is Goodfellow, Bengio et al. [37].

Some excellent free, online courses by researchers at major universities (e.g., edX,[3] MIT OpenCourseWare[4]) and corporations (e.g., Microsoft,[5] IBM,[6] Amazon,[7] DeepLearning.AI[8]) as well as online courses for a fee (e.g., Udemy,[9] Coursera,[10]· DataCamp[11]) may also be helpful.

This monograph is organized as follows. The remaining sections of this chapter will begin with a review of linear and polynomial approximation to draw analogies with very simple neural networks. Next, *neurons*, the most fundamental units of neural networks, will be used to introduce basic terminology and concepts. Then, one of the simplest examples of an neural network, the *multi-layer perception* (MLP), will be used to show why massive volumes of data are required to construct a high-quality model with predictive capability and low prediction error. The chapter concludes with a discussion of common methods used to augment data sets and some scenarios in which they can fail to solve issues with inadequate data volume. In general, synthetic data augmentation is not possible when the data set is extremely small (since statistical analysis is not possible) or when the data is highly skewed, with very few points from the minority set.

Chapter 2 introduces the notion of *multiparty computation*, in which three or more parties pool together their data to perform data analysis. We discuss simple multiparty computing models for training AI models and many issues (including legal, business, and technical issues) that circumvent their real-world deployment.

Chapter 3 introduces *edge computing*, a relatively new computing paradigm in which devices that collect data (e.g., smart phones, smart home devices, digital assistants) process and conduct analyses of the data on the device, itself. Then, it discusses the prevalence of edge computing in consumer devices used in our daily lives.

Chapter 4 introduces a specialized type of multiparty computation, known as *federated learning*, that overcomes privacy and security issues with multiparty computation by using an edge computing-like approach to enable the training of an AI

---

[3] edx.org.

[4] ocw.mit.edu.

[5] learn.microsoft.com/en-us/training.

[6] www.ibm.com/training.

[7] aws.amazon.com/training.

[8] www.deeplearning.ai/short-courses.

[9] www.udemy.com.

[10] www.coursera.org.

[11] www.datacamp.com.

model using information extracted from real-world data from multiple sources. Raw data is never shared with external parties.

Chapter 5, the final chapter, presents two examples of challenges faced by engineers during real-world deployment of federated learning and some solutions: (1) *data leakage*, i.e., reconstruction of the raw data from the AI model and results from federated learning and (2) *data poisoning*, i.e., modifying data for training an AI model to intentionally cause harm by causing the AI model to output incorrect or misleading predictions. There are countless challenges that test secure computing frameworks. The two that are presented are intended to illustrate some of the difficulties associated with introducing a new paradigm.

It is my hope that this monograph will be a friendly introduction to AI and federated learning for mathematicians and will help them communicate more effectively in collaborations with security researchers and practitioners with an engineering or computer science background.

1.2 Neural Networks: A Quick Tutorial/Review

The goal of designing neural networks is to accurately predict future outcomes based on massive troves of data from past events. Since neural network design is similar to the problem of linear regression or curve fitting, we begin with a review of polynomial approximation for a given set of points $\{x_i, y_i\}_{i=1,2,\ldots,n}$. Next, the most fundamental unit of neural networks, the *neuron*, will be presented. Finally, one of the simplest types of neural networks, the multi-layer perceptron (MLP), will be introduced. Analogies and differences between the methods will be discussed where appropriate.

1.2.1 Polynomial Regression

The goal of polynomial regression is to use a set of data points $\{x_i, y_i\}$; $i = 1, 2, \ldots, n$, to develop a mathematical model, in the form of a polynomial

$$y = a_0 + a_1 + a_2x^2 + \cdots + a_mx^m \tag{1.1}$$

that can accurately predict the value $\tilde{y}$ for any future input $\tilde{x}$. Given a set of points $\{x_i, y_i\}$; $i = 1, 2, \ldots, n$, on the xy-plane that appear to lie on a curve, one usually begins with a linear approximation, then tries higher order polynomials, and if they all fail, perhaps splines [69].

The goal of the simplest model, linear regression, is to find a line, $y = ax + b$, that is the "best fit" for the data[12,13] [65]. More precisely, define an error (or *loss*)

---

[12] Wikipedia, Simple linear regression: en.wikipedia.org/wiki/Simple_linear_regression. Accessed August 29, 2024.

[13] WikiLectures, Correlation and regression: www.wikilectures.eu/w/Correlation_and_Regression. Accessed 29 August 2024.

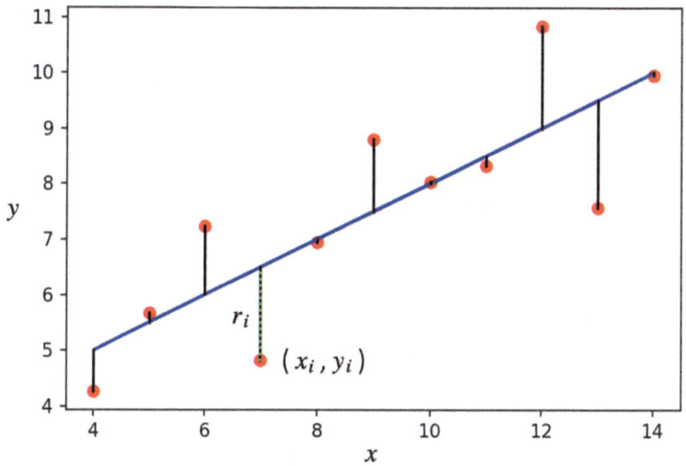

Fig. 1.1 Linear regression with least squares method. The goal of linear regression is to find a line $y = ax + b$ that is the "best fit" for the data. Define an *error* (or *loss*) function, then find the location of its minimum value, i.e., find the slope a and y-intercept b that minimize the error. For example, in the least squares method, the error function E_{linear} is defined as the sum of the squares of *residuals* r_i

function, then find the location of its minimum value, i.e., find the slope a and y-intercept b that minimize the error. For example, in the least squares method, the error function E_{linear} is defined as the sum of the squares of *residuals* r_i, as shown in Fig. 1.1:

$$r_i = y_i - (ax_i + b). \tag{1.2}$$

In the case of linear approximation, the error function is

$$E_{\text{linear}} = \sum_{i=1}^{n} r_i^2 = \sum_{i=1}^{n} [y_i - (ax_i + b)]^2. \tag{1.3}$$

For the linear least squares problem, there is a closed-form solution for the slope a and intercept b. When the mean of the data points, $\overline{x}$ and $\overline{y}$, are known (i.e., do not need to be computed), then:

$$a = \overline{y} - b\overline{x} \qquad \text{and} \qquad b = \frac{\sum_{i=1}^{n}(x_i - \overline{x})(y_i - \overline{y})}{\sum_{i=1}^{n}(x_i - \overline{x})^2}, \tag{1.4}$$

where

$$\overline{x} = \sum_i \{x\}_{i=1}^n \qquad \text{and} \qquad \overline{y} = \sum_i \{y\}_{i=1}^n. \tag{1.5}$$

If the means $\overline{x}$ and $\overline{y}$ are not known (i.e., need to be calculated), then it is more efficient to use the stand-alone equations for a and b:

$$a = \frac{\sum_{i=1}^n y_i \sum_{i=1}^n x_i^2 - \sum_{i=1}^n x_i \sum_{i=1}^n x_i y_i}{n \sum_{i=1}^n x_i^2 - \left(\sum_{i=1}^n x_i\right)^2} \tag{1.6}$$

and

$$b = \frac{n \sum_{i=1}^n x_i y_i - \sum_{i=1}^n x_i \sum_{i=1}^n y_i}{n \sum_{i=1}^n x_i^2 - \left(\sum_{i=1}^n x_i\right)^2}. \tag{1.7}$$

If the data points appear to lie on a quadratic curve rather than a straight line, we may be better off designing a quadratic approximation $y = ax^2 + bx + c$. Again, we will need to define an error function. For a least squares approximation, the residual r_i and error function $E_{\text{quadratic}}$ are

$$r_i = y_i - (ax_i^2 + bx_i + c) \tag{1.8}$$

and

$$E_{\text{quadratic}} = \sum_{i=1}^n r_i^2 = \sum_{i=1}^n [y_i - (ax_i^2 + bx_i + c)]^2. \tag{1.9}$$

To find the "best fit" quadratic curve that minimizes the error, use the closed-form solution to compute the values for a, b, and c that minimize the error $E_{\text{quadratic}}$.

Polynomial approximation is only useful in limited contexts. To approximate a more general, multivariate, convex function f, an efficient and viable option is to use the *gradient descent algorithm* to find parameters of the function that minimize the error in the approximation. Gradient descent is an iterative method that repeats the following steps until the derivative is close to zero or, more precisely, within a small, user-specified number $\epsilon > 0$ that is close to zero.

When the volume of data is massive or the neural network is large, the gradient descent algorithm is impractical since the error must be computed for each point in the data set before updating the weights of the AI model undergoing training. In this case, a variant of the gradient descent algorithm, known as *mini-batch gradient descent*, is used. In the mini-batch version, the training data set is divided into mini-batches (typically consisting of at most 50 data points), and only one mini-batch at a time is input into the gradient descent algorithm to determine the new position.

Some AI engineers favor the use of *stochastic gradient descent*, an extreme case of mini-batch gradient descent, in which each mini-batch consists of only one data point.

Gradient Descent Algorithm (for a function f):

$$p_{n+1} = p_n - \eta \nabla f(p_n). \tag{1.10}$$

1. Select a *starting position*, p_n.
2. Compute the gradient (numerical partial derivative) at the starting position.
3. If | gradient | $\geq \epsilon$, then:
   - Move a small step in the negative (opposite) direction of the gradient.
   - Set your new position p_{n+1} to be your new *starting position* and go to step 2.
4. If | gradient | $< \epsilon$, then the current position is a local minimum.

Note: $\eta \in [0, 1]$ is known as the *learning rate* of the algorithm. When η is very small, the stepsize is very small, so convergence to the minimum is very slow. When η is close to one, the stepsize is large, so convergence is fast, but there is a danger of overshooting the minimum.

The gradient descent algorithm is a general approach for finding the minimum (or local minima) of the error of an approximation of a function when there is no closed-form solution or no simple method for solving for extrema. It can be a valuable tool for training neural networks to minimize the error in prediction.

These examples from polynomial approximation include ideas that are also used in designing, testing, and deploying neural network models. In the outline of the procedures, we show that the terms *polynomial approximation* and *neural network model* (or *AI model*) are interchangeable.

In designing, testing and deploying {*polynomial approximations* -or- *neural network models*} we must:

- Make an "educated guess" about the {*polynomial degree* -or- *neural network model*} that will yield high-quality predictions to the solution for the problem at hand.
- Find the "best fit" {*polynomial* -or- *neural network model*}, i.e., the {*coefficients* -or- *weights*} that minimize the total error for data that is used for constructing (or training) the model, via the following steps:
  - For each data point, define the residual as the difference between the expected and actual outputs.

- Define an error function (e.g., sum of the absolute values of the residuals, sum of the squares of the residuals, ...).
- Find the set of values for the {*coefficients* -or- *weights*} in the candidate {*polynomial* -or- *neural network model*} that minimizes the error using a reliable algorithm, e.g., an accelerated variation of gradient descent, an optimizer specifically designed for minimizing the error in neural networks, ...
- Evaluate the {*polynomial* -or- *neural network model*} by computing the error associated with the training data, i.e., data that is used to construct the approximation.

- Avoid overfitting the {*polynomial* -or- *neural network model*} with the training data, i.e., don't use an {*unnecessarily high degree polynomial* -or- *overly complicated neural network model*} to get a near-perfect fit.
- To test for overfitting, evaluate the {*polynomial approximation* -or- *neural network*} model using *test data*, i.e., data that was not used in constructing the model. If the error is about the same as the error for *training data*, i.e., data used in construction of the model, overfitting is unlikely.
- Try several promising {*polynomial approximation* -or- *neural network*} models and select one with low error that is simple to deploy.

In the next subsection, we take a look at the mathematical model for a *neuron*, the fundamental building block of neural networks, as a prelude to introducing a full-fledged neural network.

1.2.2 Neurons: The Fundamental Unit of Neural Networks

The most basic unit of neural networks is a *neuron*, which derives its name and design from the biological neuron, a fundamental unit of the human brain and nervous system. Human neurons receive input as sensory information from the world. Then they integrate the sensory signals to decide if and where to pass on the information, i.e., which cell or part of the human body. This subsection presents a very simple example of a neuron to review basic concepts and terminology used by practitioners, such as "training an AI (or neural network) model", "activation functions", and "AI (or neural network) model weights".

A very simple mathematical model of a *neuron* (shown in Fig. 1.2) receives m-dimensional input $x = \{x_0, x_1, x_2, \ldots, x_m\}$ and produces an output $\{y\}$. A neuron processes the input and computes the output via the following steps:

- Multiply each component x_i of the input with its corresponding *synaptic weight* (or *weight*) w_i and sum the products. Let z denote this *weighted sum of the input* (or *net input*):

$$z = \sum_{i=0}^{m} w_i \cdot x_i. \tag{1.11}$$

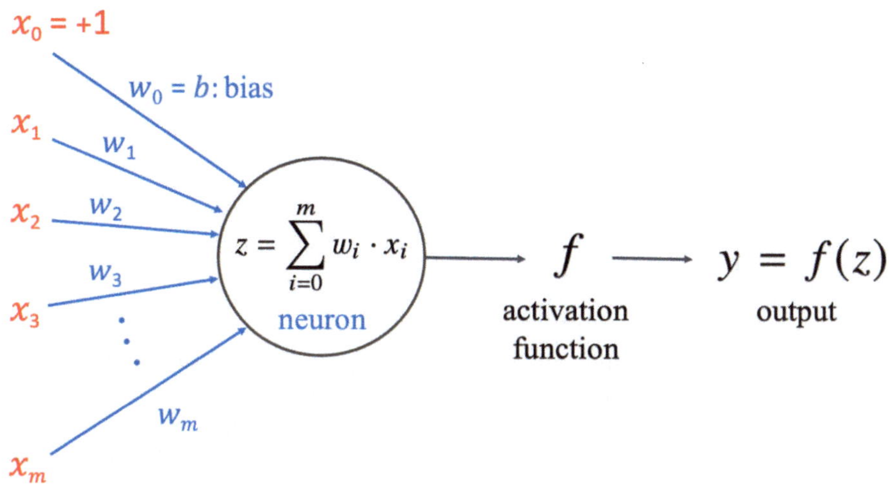

Fig. 1.2 Artificial neuron. A neuron receives input $x = \{x_0, x_1, x_2, \ldots, x_m\}$ and produces an output y via the following steps. It multiplies each component x_i of the input with the corresponding weight w_i and sums the products. Let $z = \sum_{i=0}^{m} w_i \cdot x_i$. By convention, the input variable $x_0 = 1$ serves as a constant, and the weight w_0 is known as the *bias*. The purpose of the bias is similar to the variable b, the y-intercept in linear regression. To add non-linearity to the processing, a user-specified *activation function* $f(z)$ maps input z to the final result, $y = f(z)$

- By convention, the input variable $x_0 = 1$ serves as a constant, and the weight w_0 is known as the *bias*. Its purpose is similar to the variable b, for the y-intercept in linear regression (described in Sect. 1.2.1).
- Up to this point, processing of the input is linear, which (like linear regression) may not be an appropriate representation. To add non-linearity to the processing, apply a user-specified *activation function* $f(z)$, which takes z as its input and outputs a value $f(z)$.
- Some examples of activation functions used in practice are

Binary step function: $$f(z) = \begin{cases} 0 & \text{for } z < 0 \\ 1 & \text{for } z \geq 0 \end{cases}$$

Bipolar function: $$f(z) = \begin{cases} -1 & \text{for } z < 0 \\ +1 & \text{for } z \geq 0 \end{cases}$$

Sigmoid function: $$f(z) = \frac{1}{1 + e^{-z}}$$

Hyperbolic tangent: $$f(z) = \frac{e^z - e^{-z}}{e^z + e^{-z}}$$

Rectified linear unit (ReLU): $$f(z) = \max(0, z)$$

Leaky ReLU: $$f(z) = \begin{cases} az & \text{for} \quad z < 0\,;\; 0 < a \le 0.2 \\ z & \text{for} \quad z \ge 0 \end{cases}$$

Plots of the activation function are shown in Fig. 1.3. In addition to adding non-linearity, some are convenient for converting the output $y = f(z)$ to a probability, with value in the unit interval, i.e., $0 \le f(z) \le 1$. If the output is expected to be binary (i.e., $y = 0$ or $y = 1$), the output can be converted from a real number in the unit interval to binary form by setting:

$$y = \begin{cases} 0 & \text{for} \quad 0 \le f(z) < 0.5 \\ 1 & \text{for} \quad 0.5 \le f(z) \le 1. \end{cases}$$

To summarize this subsection thus far (Fig. 1.2): Given a neuron, i.e.:

- the number of input parameters m,
- weights $w = \{w_0, w_1, w_2, \ldots, w_m\}$, and
- activation function f,

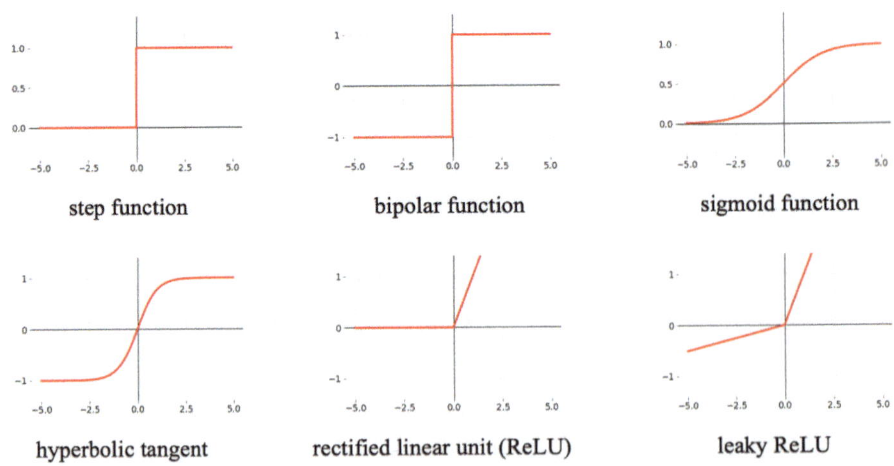

Fig. 1.3 Activation functions. Examples of activation functions used in practice for neural networks

the neuron processes input data $x = \{x_0, x_1, x_2, \ldots, x_m\}$ to generate an output $y = f(z)$, where $z = \sum_{i=0}^{m} x_i \cdot w_i$ and $f(z)$ is an activation function.

We know the number of parameters of the input data, and we can select an activation function following the example of a similar problem that has been successfully modeled by neural networks for practical use. The remaining pressing question is: *Do we know the values for the weights* $w = \{w_0, w_1, w_2, \ldots, w_m\}$? The answer is: *We don't.* The work of AI scientists is to find the values for the weights that are a "best fit" for a set of *training data*, i.e., a large set of data that consists of both the input $x = \{x_0, x_1, x_2, \ldots, x_m\}$ and its corresponding output y.

In linear regression, the input was a set of points on the two-dimensional plane, and the goal was to find the slope and y-intercept of a line that minimizes the total "error" or "distance" between the data points and the line. In the problem involving design of a neuron, we need to find the weights $w = \{w_0, w_1, w_2, \ldots, w_m\}$ that minimize the total error (sum of the errors) of the training data. As with polynomial approximation, a substantial volume of high-quality data (low noise, low error) is needed to solve the problem.

Since neural processing was purposely designed to be non-linear, the problem is significantly more complicated than polynomial approximation. Computers must numerically compute estimates for the weights $w = \{w_0, w_1, w_2, \ldots, w_m\}$ via an iterative process that converges to the "best fit" solution by moving toward the minimum of the error using a optimizer such as gradient descent (GD), mini-batch gradient descent, stochastic gradient descent (SGD) (introduced in Sect. 1.2.1) or one of its variants such as *adaptive moment estimation* (Adam)[14] [48]. This computational process of converging toward the minimum of the error is known as "*training a neural network model*" or "*training an AI model*".

1.2.3 Multi-layer Perceptrons: A Simple Example

This subsection walks through an example of one of the simplest neural network models, a *feed-forward multi-layer perceptron* (MLP), shown in Fig. 1.4. This very simple example can explain why so much data is needed to train neural networks that will be deployed in the real world.

The underlying ideas behind multi-layer perceptrons (and all neural networks) were inspired by the human neural processing system, which consists of many neurons working together to process information. Multi-layer perceptrons are built from combining neurons, enabling them to receive inputs, and transmitting the information (sometimes in modified form) to the brain and between other neurons. This small, toy model will show how an input is processed and why a tremendous amount of data is needed to "*train*" neural network/AI models, i.e., find the weights between all pairs of nodes that are connected.

[14] Adam has become a very popular algorithm to use for gradient descent during initial implementations of several types of neural networks to test their viability for real-world deployment.

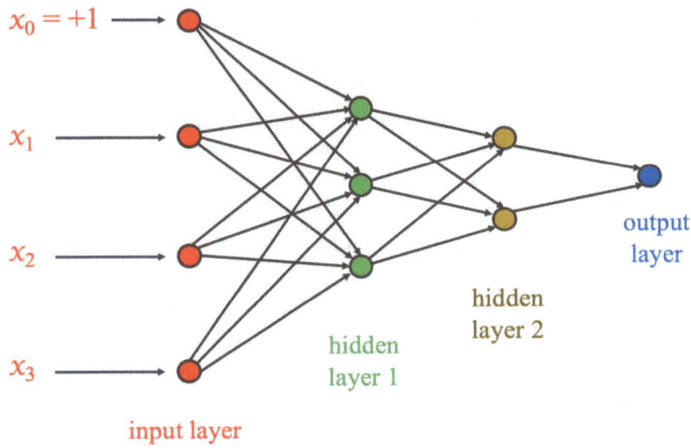

Fig. 1.4 Feed-forward multi-layer perception. Simple example of a neural network

Given a set of training data (a very large set of input data and the result), the goal is to find the "optimal" combination of weights and activation functions for all links that connect one node with another in the next layer in Fig. 1.4. Here "optimal" denotes the set of weights and activation functions that yield the lowest total error for the training data. The training process proceeds as follows:

- Since the weights are unknown at the very beginning of the training process, they are all set to one (or some random, non-zero value).
- For each input data object $x_i = \{x_{i,0},\ x_{i,1},\ x_{i,2},\ x_{i,3}\}$, process it to obtain an output $\hat{y}_i$ by
  - multiplying each $x_{i,j}$ by the weights associated with the edges connecting them to hidden layer 1;
  - summing the products of the $x_{i,j}$ and weights at each node in hidden layer 1;
  - passing the sum through an activation function to find the final output values at nodes in hidden layer 1;
  - carrying out the analogous process (for each node in hidden layer 1) to compute values for nodes in hidden layer 2;
  - carrying out the analogous process (for each node in hidden layer 2) to compute values for the output $\hat{y}_i$;
  - computing $\Delta y_i = (y_i - \hat{y}_i)$, the difference between the computed output value $\hat{y}_i$ and the actual output value y_i in the training data;
  - computing the total error for the iteration using a well-defined error function, e.g., sum of squares $\sum(\Delta y_i)^2$ or sum of absolute values $\sum |\Delta y_i|$.
- If the error is larger than an acceptable value, use an optimizer developed for neural network training (e.g., stochastic gradient descent, mini-batch gradient descent,

adaptive moment estimation, or Adam[15]) to adjust the weights to decrease the total error for the training set, and repeat the iterative loop above until:

- the error is less than or equal to an acceptable value; or
- the error is only decreasing by a very small amount between two iterations (below a pre-set threshold value); or
- a pre-set number of iterations (normally a few hundred) is completed.

The training problem for even this small, toy neural network is very complicated and normally requires hundreds of iterations of the computations before the weights converge to a neural network model with an error that is acceptable. Most neural networks deployed for real-world use consist of hundreds of input parameters, at least several hidden layers, and hundreds of processing nodes. Training these neural networks for deployment requires a massive volume of data. Imagine the volume of data required for polynomial regression, then for approximations of a surface in multidimensional space. Training neural networks is even more complicated and requires significantly more data to find weights and activation functions that yield a model of practical value, i.e., one that will yield useful predictions.

Another concern for real-world deployment of AI systems is finding and fine-tuning an appropriate neural network model. Engineers need to try out and compare results from several different types of neural network models before selecting the most promising one. Training costs for multiple models can become very high very quickly. For neural network models that will be used frequently in production, fine-tuning will improve predictive capability, so several variations of the selected model will also need to be trained for comparison.

Finding, labeling, or generating sufficient volumes of high-quality data with input and output values, then storing them for future use (e.g., for updating the neural network as more data becomes available) can become very expensive. The expense, logistics, and legal compliancy issues involved should be taken into account when considering the introduction of an AI system for real-world use. For scenarios that frequently undergo a lot of change or have seasonality issues, costs for updating the model should also be considered.

Training neural networks is associated with another major problem. The amount of computing power, electricity, and water required to train a prospective model can be enormous [9, 30, 35, 40, 47, 55]. In real-world deployments, this cost should be multiplied at least by several times since several different types of models are usually tried out and compared for accuracy, robustness, speed, etc. The most promising model is further refined and fine-tuned, using even more computing power and electricity.

Recently, the unconstrained consumption of natural resources for training and running AI models has become a major environmental, political, and business issue. Its impact is being followed closely by the mainstream media [30, 57]. In September of 2024, Google was forced to reconsider plans to build a massive new data center in

[15] Stochastic gradient descent, Wikipedia: en.wikipedia.org/wiki/Stochastic_gradient_descent.

Chile when environmental concerns, including the annual need for millions of liters of water usage for cooling, were cited as a major problem for a country suffering through a major drought.[16] Opposition from environmentalists and regions targeted for hosting new data centers is expected to escalate unless new technologies or approaches can be found to reduce environmental damage.

Unfortunately, the data volume and power consumption problems have become even more acute following the development and proliferation of generative AI. Early AI systems, such as multi-layer perceptrons, focus on predicting or classifying input [60]. Generative AI must generate text, images, audio, and video by identifying patterns in training data and combining them following instructions (known as prompts) from the user. The volume of data and computational power needed for training generative AI systems are orders of magnitude larger than earlier AI systems and are major impetuses for cloud service providers to further accelerate their already controversial pace of data center construction, electrical generation facilities, and water consumption.

1.3 Increasing Data Volume

As mentioned earlier, many decision-makers for institutions with digital data would like to use AI to predict the future, find outliers, discover illegal activities, detect rare conditions and diseases, etc. However, the institution may have insufficient data to train a robust and reliable AI model. Depending on the application scenario, two approaches that may help overcome this impasse are data augmentation and data synthesis. *Data augmentation* methods start with real data, then generate new data by making minor changes. *Data synthesis* methods generate new data without relying on a real data set. This section examines techniques for data augmentation and data synthesis and limitations on their usefulness for increasing the volume of data for training AI models for real-world deployment.

1.3.1 Data Augmentation

Data augmentation can be defined as methods for artificially generating new data, based on properties of existing real-world data, to increase data volume[17,18] [19, 20, 36, 66, 80, 95]. When constructing augmentation methods to

[16] Reuters (September 18, 2024) Google takes Chile data center plans back to square one on environmental concerns: www.reuters.com/technology/google-takes-chile-data-center-plans-back-square-one-environmental-concerns-2024-09-17/. Cited 18 September 2024.

[17] Scikit-image: Geometrical transformations of images, online tutorial: scikit-image.org/docs/stable/user_guide/geometrical_transform.html.

[18] Zhou, Y., Guo, C., Wang, X., Chang, Y., Wu, Y., Papers and resources on data augmentation using large models: github.com/MLGroup-JLU/LLM-data-aug-survey.

modify real-world data, an important property to verify is *class label preservation* [18]. Shorten and Khoshgoftaar [77] examined the probability of an augmentation method to preserve the class label, and defined it as the *safety of an augmentation method*.

Data augmentation methods typically start with a small real-world data set, then add noise and/or some perturbation. Some examples and references are

- duplicating and appending the data with/without random noise;
- supplementing with similar type of open data;
- changing the speed, pitch, ... in part or all of an audio data segment [51];
- data warping, geometric transformations, random flipping, cropping, mixing, and random erasing of images[19] [26, 33, 94];
- augmenting text data [7, 29, 36, 70];
- augmenting time-series data [88];
- augmenting medical image data [46].

Duplicating data and adding noise is different for different domains. For text data, adding common typos of words or replacing words with their synonyms are some common methods. For audio data, noise may be added, or the speed or pitch may be adjusted. For numerical data, Gaussian noise may be inserted into the data. Scientists believe that astute addition of noise can improve training results since it may prevent AI models from memorizing training examples[20] [56, 72]. Before adding noise, parameters in the data should be normalized (or rescaled) to make sure that they are in the same range so that the noise has the same effect on all parameters. The level of appropriate noise should be systematically tested to make sure it is neither too much or too little.

Some methods were developed to augment a specific data type, such as text, audio, or images, while others can be applied to a wider range of data types. Research on data augmentation has been made and continues to make tremendous strides. Since it is beyond the scope of this monograph to review current or future breakthroughs in this research area, we refer interested readers to surveys on the topic [7, 11, 29, 42, 44, 46, 63, 77, 88, 91].

Imbalanced data sets are data sets in which the frequency of the majority class is much larger than that of the minority class(es). Data augmentation for highly imbalanced data sets may be necessary even when there is a massive volume of data if the smaller subset is extremely rare since there may be insufficient data from the smaller subset for statistical analysis. For example, in banking transactions, fraudulent cases are far less than one in a million. In medicine, many doctors and hospitals will never encounter a patient with an extremely rare disease.

---

[19] Jason Brownlee (July 5, 2019) How to configure image data augmentation in Keras: machinelearningmastery.com/how-to-configure-image-data-augmentation-when-training-deep-learning-neural-networks.

[20] Jason Brownlee (6 August 2019) Train neural networks with noise to reduce overfitting: machinelearningmastery.com/train-neural-networks-with-noise-to-reduce-overfitting.

For some very large imbalanced data sets, depending on the application, it may be possible to reduce the size of the larger subset by randomized sampling [64] and apply data augmentation to the smaller subset to construct a relatively well-balanced data set for training. Sampling the large data set might result in loss of some information, so it is critical to assess whether it is important to retain information on outliers or small clusters for the intended application. Approaches to data augmentation for highly imbalanced data sets are discussed in detail in [8, 13, 21, 22, 43, 54]. Most of the methods assume two class problems (e.g., cancerous/benign, fraudulent/normal, yes/no), but multiple class imbalance is also a common problem [86].

In closing this subsection, we note that data augmentation techniques need to be monitored closely to prevent adversarial examples from polluting a data set that will be used for training an AI model. Some types of data poisoning attacks are off-shoots of data augmentation techniques. It is possible to make a very small change in data that causes poor training of AI models [4, 28]. A very simple example for text data is insertion of the word "*not*" in a sentence, which has the effect of negating the original text. For image data, data sets for poisoning start with the real, raw data sets, then make very small, targeted changes that are indiscernible to the human eye.[21] The field of data poisoning also encompasses data synthesis techniques and hybrids, which will be discussed in more detail in Sect. 5.3.

1.3.2 Data Synthesis

Data synthesis has been an important tool in software testing.[22] Some benefits of using synthetic data are

- It can produce as much data as needed.
- It is reliable (once the software is written) and can be tested.
- It can mitigate issues with security and privacy compliance.
- It can be designed to have specific characteristics, e.g., to be similar to customer demographics (age, sex, income, ...).
- It can produce diverse data sets that are free of bias,[23] with outliers and edge cases.

Synthetic data should not be used when: the properties of real data are unknown; the cost of producing the data is significant; or when important properties real data would be very difficult to emulate.

---

[21] Jonas Geiping, Industrial scale data poisoning website: github.com/JonasGeiping/data-poisoning.

[22] Haritha Khandabattu, Tigran Egiazarov, Alys Woodward, Ross Power, Use Synthetic Data to Improve Software Quality, Gartner Report, 24 Apr. 2024.

[23] Writing programs to synthesize unbiased, diverse data sets is not always easy. General rules may not work, as in the famous case of Google's AI Gemini image generator "depicted America's founding fathers and Nazi soldiers as Black": www.npr.org/2024/03/18/1239107313/google-races-to-find-a-solution-after-ai-generator-gemini-misses-the-mark.

Recently, the need for data for training machine learning and AI models has expanded the business prospects for data synthesis. According to a Gartner Press Release in August of 2023[24]:

> The use of generative AI to create synthetic data is one area that is rapidly growing, relieving the burden of obtaining real-world data so machine learning models can be trained effectively. By 2024, Gartner predicts 60% of data for AI will be synthetic to simulate reality, future scenarios and de-risk AI, up from 1% in 2021.

Some commonly used approaches for generating synthetic data are

- rule-based logic;
- statistical methods that emulate distributions of real data;
- generative adversarial networks (GANs)[25] [38, 41];
- variational autoencoders (VAEs)[26] [49];
- large language models (LLMs);
- diffusion models [41, 78]; and
- neural radiance fields (NeRFs) [61].

The first two approaches are traditional methods that have been used over the years. The latter six are generative methods that rely on neural networks and are relatively new. The inner workings of neural networks are not fully understood, and some highly unexpected results may be produced, so it is important for a human to check the output synthetic data. Famous examples are hallucinations output by large language models [25, 45, 59]. Users should be beware of quirks and cautionary notes for a given generative AI method and output data type.

Generative adversarial networks (GANs) were introduced by Goodfellow et al. [38]. The method trains two AI models: one that uses real data to produce synthetic data, and a second that guesses whether a data sample came from the training data or was synthesized. Training of both models continues until the quality of synthesized data improves to such a high level that the second AI model cannot differentiate between real and synthesized data. GANs tend to suffer from non-convergence and instability, but there has been some progress in addressing these problems. Also, generated data may have less diversity than may be required for some applications. Many variations and extensions of GANs have been and continue to be developed for a wide variety of applications [1, 10, 23, 39, 41, 58].

Variational autoencoders (VAEs) generate new data (e.g., images) in two steps: map each raw data object into a distribution within a latent space of lower dimension. The latent space should be a subspace that keeps many of the important properties of the data intact. Next, map the latent space back to the higher-dimensional input

[24] Gartner Press Release Newsroom: www.gartner.com/en/newsroom/press-releases/2023-08-01-gartner-identifies-top-trends-shaping-future-of-data-science-and-machine-learning.

[25] Lilian Weng, From GAN to WGAN, August 20, 2017: lilianweng.github.io/posts/2017-08-20-gan. Cited 18 July 2024.

[26] Lilian Weng, From Autoencoder to Beta-VAE, August 12, 2018: lilianweng.github.io/posts/https://lilianweng.github.io/posts/2018-08-12-vae/.

space. During expansion, knowledge of the statistical properties of a raw data set is used to synthesize new data. For image data, VAEs tend to produce lower-quality, fuzzy images than GANs.

Large language models (LLMs) have been receiving a lot of attention following the introduction of ChatGPT to the general public. These generative AI models are trained on massive volumes of text data, and they use information from the context of words and phrases and their relationships with other words and phrases to produce test output. Some recent surveys in this area are[27] [3, 12, 15, 16, 52, 62, 67, 68, 75, 84, 93]. This field of research is moving very rapidly, so readers should look for even more recent works to understand the current status.

Diffusion models were introduced by Sohl-Dickstein et al. [78]. They also consist of two steps. In the first, *forward diffusion* step, noise is gradually added to raw data. In the second *backward diffusion* step, noise is slowly deleted. Noise removal is not an inverse operation, but a process that restores structure in data. Diffusion models have been successfully used in some applications to generate training data and can be used to combine features from different classes to generate new hybrid image data [6, 73, 76, 83].

Latent diffusion [74], a variation of diffusion models, has become famous due to the release of *Stable Diffusion*,[28] a software package whose generated images were featured by mass media outlets. Stable Diffusion's release is under the Creative ML OpenRAIL-M license, which permits use, change, and redistribution of modified software. However, any derivative software must also be released under the same license and include a copy of the original Stable Diffusion license.[29] Tutorials on diffusion and latent diffusion include[30] [2].

Neural radiance fields were introduced by Mildenhall et al. [61]. They are fully connected neural networks that can generate complex, 3D scenes using a set of 2D images. Initially, they were used to generate computer graphics scenes for video games and animations in films. Neural radiance fields have been subsequently used by other researchers to generate photorealistic scenes for videos [14, 27, 32, 53, 85, 90, 92].

For readers interested in learning more about AI and generative AI, a free, online MIT course on Deep Learning[31] is updated annually and follows the fast-paced development of the field. The 2024 course covers generative AI systems, including sample code for hands-on study.[32] Although background in AI is not a prerequisite,

[27] Efficient Large Language Models: A Survey, a github page with links to papers, code and website with new developments by some authors of [84]: github.com/AIoT-MLSys-Lab/Efficient-LLMs-Survey.

[28] Stable Diffusion Public Release: stability.ai/news/stable-diffusion-public-release.

[29] Stable Diffusion License: huggingface.co/spaces/CompVis/stable-diffusion-license.

[30] How does Stable Diffusion work?—Latent Diffusion Models EXPLAINED, AI Coffee Break with Letitia: www.youtube.com/watch?v=J87hffSMB60.

[31] Amini, Alexander, 2024 MIT Introduction to Deep Learning 6.S191: introtodeeplearning.com/.

[32] Amini, Alexander, 2024 MIT Introduction to Deep Learning 6.S191, github site for hands-on labs: github.com/aamini/introtodeeplearning/tree/master.

completion of a more slow-paced elementary course on neural networks (e.g., on Udemy, DataCamp, Coursera) may be helpful.

Sometimes generation of synthetic data is not a viable option. Very small data sets may not have complete information to generate realistic data. Similarly, the minority class in extremely imbalanced data sets (e.g., for banking fraud detection, medical data for diagnosis) may consist of so few data points that data augmentation cannot lead to an accurate AI model with useful predictive capabilities. In the next two chapters, we discuss alternative approaches to resolve issues with small data sets with inadequate or incomplete information. The approaches require cooperation by multiple owners of small or skewed data sets to facilitate aggregation of information in their data sets to increase the volume of training data.

References

1. Aggarwal, A., Mittal, M., Battineni, G.: Generative adversarial network: an overview of theory and applications. Int. J. Inf. Manag. Data Insights **1**(1) (2021). sciencedirect.com/science/article/pii/S2667096820300045. Cited 11 March 2024
2. Alammar, J.: The illustrated stable diffusion, 4 (2022). jalammar.github.io/illustrated-stable-diffusion/. Cited 11 March 2024
3. Alammar, J., Grootendorst, M.: Hands-On Large Language Models. O'Reilly Media Inc, Sebastopol, CA, USA (2024)
4. Alzantot, M., Sharma, Y., Elgohary, A., Ho, B.J., et al.: Generating natural language adversarial examples. In: Proceedings of Empirical Methods in Natural Language Processing, pp. 2890–2896 (2020)
5. Axler, S.: Linear Algebra Done Right, 2nd edn. Springer, Berlin, Germany (1997). link.springer.com/book/10.1007/ 978-3-319-11080-6. Cited 11 March 2024
6. Azizi, S., Kornblit, S., Saharia, C., Norouzi, M., Fleet, D.J.: Synthetic data from diffusion models improves ImageNet classification. Trans Mach. Learn. Res. **10** (2023). arxiv.org/abs/2304.08466. Cited 11 March 2024
7. Bayer, M., Kaufhold, M.-A., Reuter, C.: A survey on data augmentation for text classification. ACM Comput. Surv. **55**(7), 1–39 (2022). dl.acm.org/doi/abs/10.1145/3544558. Cited 9 July 2024
8. Bellinger, C., Corizzo, R., Japkowicz, N.: Calibrated resampling for imbalanced and long-tails in deep learning. in: Discovery Science, Lecture Notes in Computer Science, vol. 12986, pp. 242–252. Springer, Berlin, Germany (2021). arxiv.org/abs/2012.02312. Cited 8 July 2024
9. Berreby, D.: As use of A.I. soars, so does the energy and water it requires. Yale Environ. **360** (2024). e360.yale.edu/features/artificial-intelligence-climate-energy-emissions. Cited 26 April 2024
10. Biswas, A., et al.: Generative adversarial networks for data augmentation. In: Zheng, B., Andrei, S., Sarker, Md.K., Datta, K., Guptain (eds.) Data Driven Approaches on Medical Imaging, pp. 159–177. Springer, Heidelberg, Germany (2023). arxiv.org/abs/2306.02019v2. Cited 14 March 2024
11. Cauli, N., Recupero, D.R.: Survey on videos data augmentation for deep learning models. Future Internet **14**(3), 93 (2022). www.mdpi.com/1999-5903/14/3/93. Cited 14 March 2024
12. Chang, Yupeng, et al.: A survey on evaluation of large language models. ACM Trans. Intell. Syst. Technol. **15**(3), 1–45 (2024). dl.acm.org/doi/full/10.1145/3641289. Cited 21 July 2024
13. Chawla, N.V., Bowyer, K.W., Hall, L.O., Kegelmeyer, W.P.: Smote: synthetic minority oversampling technique. J. Artif. Intell. Res. **16**, 321–357 (2022)

14. Cheng, W., Xu, S., Piao, J., Qian, C., et al.: Generalizable neural performer: learning robust radiance fields for human novel view synthesis (2022). arxiv.org/abs/2204.11798. Cited 26 March 2024
15. Chockalingam, A., Patel, A., Verma, S., Yeung, T.: A Beginner's Guide to Large Language Models. NVIDA Corp., Santa Clara, CA, USA (2023). resources.nvidia.com/en-us-large-language-model-ebooks/llm-ebook-part1. Cited 20 July 2024
16. Chockalingam, A., Patel, A., Verma, S., Yeung, T.: How LLMs are Unlocking New Opportunities for Enterprises. NVIDA Corp., Santa Clara, CA, USA (2023). resources.nvidia.com/en-us-large-language-model-ebooks/llm-ebook-part1. Cited 20 July 2024
17. Chollet, F.: Deep Learning with Python, 2nd edn. Manning, Shelter Island, NY, USA (2021)
18. Coulombe, C.: Text data augmentation made simple by leveraging NLP cloud APIs (2018). arxiv.org/abs/1812.04718. Cited 10 July 2024
19. Cubuk, E.D., Zoph, B., Mane, D., Vasudevan, V., Le, Q.V.: AutoAugment: learning augmentation strategies from data. In: Proceedings of the IEEE Computer Vision and Pattern Recognition Conference on (CVPR), pp. 113–123 (2019)
20. Cubuk, E.D., Zoph, B., Shlens, J., Le, Q.V.: Randaugment: practical automated data augmentation with a reduced search space. In: Proceedings of the IEEE Computer Vision and Pattern Recognition Workshops (CVPRW), pp. 702–703 (2020)
21. Dablain, D., Bellinger, C., Krawczyk, B., Chawla, N.: Efficient augmentation for imbalanced deep learning. In: Proceedings of the IEEE International Conference on Data Engineering (ICDE), pp. 1433–1446 (2023). arxiv.org/abs/2207.06080. Cited 11 March 2024
22. Dablain, D., Chawla, N.: Towards understanding how data augmentation works with imbalanced data (2023). arxiv:2304.05895. Cited 6 July 2024
23. Dash, A., Ye, J., Wang, G.: A review of generative adversarial networks (GANs) and its applications in a wide variety of disciplines—From medical to remote sensing. IEEE Access (2023). ieeexplore.ieee.org/stamp/stamp.jsp?arnumber=10372211. Cited 11 March 2024
24. Deisenroth, M.P.: Mathematics for Machine Learning. Cambridge University Press, Cambridge, UK (2020)
25. Diamond, S.: A.I. Chatbots, Hens and Humans Can All 'Hallucinate', NY Times (2023). www.nytimes.com/2023/12/17/insider/ai-chatbots-humans-hallucinate.html. Cited 14 October 2024
26. Dilmegani, C.: Top data augmentation techniques: ultimate guide for 2024. AI Multiple Research website (2024). research.aimultiple.com/data-augmentation-techniques. Cited 14 March 2024
27. Di Nucci, D., Simoni, A., Tomei, M., Ciuffreda, et al.: CarPatch: a synthetic benchmark for radiance field evaluation on vehicle components. In: Foresti, G.L., Fusiello, A., Hancock, E. (eds.) Image Analysis and Processing. Proceedings of the International Conference on Image Analysis and Processing (ICIAP), Lecture Notes in Computer Science, vol. 14234, pp. 99–110. Springer, Cham, Switzerland (2023). arxiv.org/abs/2307.12718. Cited 26 March 2024
28. Ebrahimi, J., Rao, A., Lowd, D., Dou, D.: Hotflip: white-box adversarial examples for text classification. In: Proceedings of Annual Meeting of the Association for Computational Linguistics, vol. 2, pp. 31–36 (2018)
29. Feng, S.Y., Gangal, V., Wei, J., Chandar, S., et al.: A survey of data augmentation approaches for NLP. In: Findings of the Association for Computational Linguistics, pp. 968–988 (2021)
30. Font, J.: Working at full power: data centers in the era of AI, Forbes Technology Council (2023). www.forbes.com/sites/forbestechcouncil/2023/10/30/working-at-full-power-data-centers-in-the-era-of-ai/?sh=4571249a3ab0. Cited 26 April 2024
31. Fortune Business Insights website.: Big data analytics market size, share & industry analysis (2024). www.fortunebusinessinsights.com/big-data-analytics-market-106179. Cited 11 March 2024
32. Ge, Y., Behl, H., Xu, J., Gunasekar, S., et al.: Neural-Sim: learning to generate training data with NeRF (2022). arxiv.org/abs/2207.11368. Cited 16 March 2024
33. Glasbey, C.A., Mardia, K.V.: A review of image-warping methods. J. Appl. Stat. **25**(2), 155–171 (1998). www.math.chalmers.se/~rudemo/Papers/GlasbeyMardia98.pdf. Cited 26 March 2024

34. Glassner, A.: Deep Learning: a Visual Approach. No Starch Press, Burlingame, CA, USA (2021)
35. Goldman Sachs.: AI is poised to drive 160% increase in data center power demand (2024). www.goldmansachs.com/intelligence/pages/AI-poised-to-drive-160-increase-in-power-demand.html. Cited 8 July 2024
36. Goodfellow, I.: Reddit MachineLearning—Generative adversarial networks for text (2015). bit.ly/2ES5Elq. Cited 11 March 2024
37. Goodfellow, I., Bengio, Y., Courville, A.: Deep Learning. MIT Press, Cambridge, MA, USA (2016)
38. Goodfellow, I., Pouget-Abadie, J., Mirza, M., Xu, B., et al.: Generative adversarial networks. Commun. ACM **63**(11), 139–144 (2020). dl.acm.org/doi/10.1145/3422622. Cited 11 March 2024
39. Gonog, L., Zhou, Y.: A review: generative adversarial networks. In: Proceedings of the IEEE Conference on Industrial Electronics and Applications (ICIEA) (2019). ieeexplore.ieee.org/document/8833686. Cited 16 March 2024
40. Guerrini, F.: AI's unsustainable water use: how tech giants contribute to global water shortages. Forbes (2023). www.forbes.com/sites/federicoguerrini/2023/04/14/ais-unsustainable-water-use-how-tech-giants-contribute-to-global-water-shortages. Cited 8 July 2024
41. Gui, J., Sun, Z., Wen, Y., Tao, D., Ye, J.: A review on generative adversarial networks: algorithms, theory, and applications. IEEE Trans. Knowl. Data Eng. **35**(4), 3313–3332 (2023). ieeexplore.ieee.org/document/5288526, arxiv.org/abs/2001.06937. Cited 11 March 2024
42. Haba, D.: Data Augmentation with Python. Packt Publishing, Birmingham, UK (2023)
43. He, H., Bai, Y., Garcia, E. A., Li, S.: Adasyn: adaptive synthetic sampling approach for imbalanced learning. In: Proceedings of the IEEE International Joint Conference on Neural Networks, pp. 1322–1328 (2008). www.researchgate.net/publication/224330873_ADASYN_Adaptive_Synthetic_Sampling_Approach_for_Imbalanced_Learning. Cited 8 July 2024
44. Hernández-García, A.: Data augmentation and image understanding. Ph.D. Thesis, Institute of Cognitive Science, University of Osnabrück, Germany (2020)
45. Ji, Z., Lee, N., Frieske, R., Yu, T. et al.: Survey of hallucination in natural language generation. ACM Comput. Surv. **55**(12), 1–38 (2022). dl.acm.org/doi/abs/10.1145/3571730. Cited 8 October 2024
46. Kebaili, A., Lapuyade-Lahorgue, J., Ruan, S.: Deep learning approaches for data augmentation in medical imaging: a review. J. Imaging **9**(4), 8 (2023). www.mdpi.com/2313-433X/9/4/81. Cited 16 March 2024
47. Kindig, B.: AI power consumption: rapidly becoming mission-critical. Forbes (2024). www.forbes.com/sites/bethkindig/2024/06/20/ai-power-consumption-rapidly-becoming-mission-critical. Cited 8 July 2024
48. Kingma, D.P., Ba, L.B.: Adam: a method for stochastic optimization. In: Proceedings of the International Conference on Learning Representations (ICLR) (2015). arxiv.org/abs/1412.6980. Cited 26 August 2024
49. Kingma, D.P., Welling, M.: Auto-encoding variational Bayes. In: Proceedings of the International Conference on Learning Representations (ICLR) (2013). arxiv.org/abs/1312.6114. Cited 16 July 2024
50. Kneusel, R.T.: Math for Deep Learning. No Starch Press, San Francisco, USA (2022)
51. Ko, T., Peddinti, V. Povey, D., Khudanpur, S.: Audio augmentation for speech recognition. In: Proceedings of Annual Conference on International Speech Communication Association, pp. 3586–3589 (2015)
52. Kublik, S., Saboo, S.: Building Innovative NLP Products Using Large Language Models. O'Reilly Media Inc, Sebastopol, CA, USA (2022)
53. Kwon, Y., Kim, D., Ceylan, D., Fuchs, H.: Neural image-based avatars: generalizable radiance fields for human avatar modeling. In: Proceedings of International Conference on Learning Representations (ICLR), Poster (2023). iclr.cc/virtual/2023/poster/10830. arxiv.org/abs/2304.04897. Cited 26 March 2024

54. Leevy, J.L., Khoshgoftaar, T. M., Bauder, R.A., Seliya, N.: A survey on addressing high-class imbalance in big data. J. Big Data **5**, 42 (2018). journalofbigdata.springeropen.com/articles/10.1186/s40537-018-0151-6. Cited 16 March 2024
55. Leffer, L.: The AI boom could use a shocking amount of electricity. Sci. Am. (2023). www.scientificamerican.com/article/the-ai-boom-could-use-a-shocking-amount-of-electricity. Cited 26 April 2024
56. Li, L., Spratling, M.: Data augmentation alone can improve adversarial training. In: Proceedings of the International Conference on Learning Representations (ICLR) (2023). arxiv.org/abs/2301.09879. Cited 12 July 2024
57. Loten, A.: AI-ready data centers are poised for fast growth. Wall Street J. (2023). www.wsj.com/articles/ai-ready-data-centers-are-poised-for-fast-growth-fadae952. Cited 12 July 2024
58. Megahed, M., Mohammed, A.: A comprehensive review of generative adversarial networks: fundamentals, applications, and challenges. WIREs Comput. Stat. (2023). wires.onlinelibrary.wiley.com/doi/full/10.1002/wics.1629. Cited 16 March 2024
59. Metz, C.: Chatbots may 'hallucinate' more often than many realize. NY Times (2023). https://www.nytimes.com/2023/11/06/technology/chatbots-hallucination-rates.html. Cited 14 October 2024
60. Meyer, D.: The cost of training AI could soon become too much to bear. Fortune (2024). fortune.com/2024/04/04/ai-training-costs-how-much-is-too-much-openai-gpt-anthropic-microsoft. Cited 26 April 2024
61. Mildenhall, B., Srinivasan, P.P., Tancik, M., Barron, J.T., et al.: NeRF: representing scenes as neural radiance fields for view synthesis. In: Proceedings of the European Conference on Computer Vision, Lecture Notes in Computer Science, vol. 12346, pp. 405–421 (2020). arxiv.org/abs/2003.08934. Cited 25 March 2024
62. Minaee, S., Mikolov, T., Nikzad, N., Chenaghlu, M., et al.: Large language models: a survey (2024). arxiv.org/abs/2402.06196. Cited 21 July 2024
63. Moisés, A.G., Pascual, I.V., González, J.J.I, Zamarreño, C.R.: Data augmentation techniques for machine learning applied to optical spectroscopy datasets in agrifood applications: comprehensive review. Sensors **23**, 8562 (2023). www.mdpi.com/1424-8220/23/20/8562. Cited 14 March 2024
64. Motwani, R., Raghavan, P.: Randomized Algorithms. Cambridge University Press, Cambridge, UK (1995)
65. Muthukrishnan.: Maths behind polynomial regression (2018). muthu.co/maths-behind-polynomial-regression/. Cited 30 August 2024
66. Nanthini, K., Sivabalaselvamani, D., Chitra, K., Gokul, P., et al.: A survey on data augmentation techniques. In: Proceedings of the International Conference on Computing Methodologies and Communication (ICCMC), pp. 913–920 (2023). ieeexplore.ieee.org/document/10084010. Cited 14 July 2024
67. Naveed, H., Khan, A. U., Qiu, S., Saib, M., et al.: A comprehensive overview of large language models (2024). arxiv.org/abs/2307.06435. Cited 20 July 2024
68. Ozdemir, S.: Quick Start Guide to Large Language Models: strategies and Best Practices for Using ChatGPT and Other LLMs. Addison-Wesley, Boston, MA, USA (2020)
69. Powell, M.J.D.: Approximation Theory and Methods. Cambridge University Press, Cambridge, UK (1981)
70. Qiu, S., Xu, B., Zhang, J., Wang, Y., et al.: EasyAug: an automatic textual data augmentation platform for classification tasks. In: Companion of the World Wide Web Conference, pp. 249–252 (2020)
71. Raff, E.: Inside Deep Learning: math, Algorithms, Models. Manning, Shelter Island, NY, USA (2023)
72. Rebuffi, S.-A., Gowal, S., Calian, D.A., Stimberg, F., et al.: Data augmentation can improve robustness. Neural Inf. Process. Syst. (2021). openreview.net/forum?id=kgVJBBThdSZ. Cited 12 July 2024

73. Ridge-Institute R& D Blog: using diffusion models to generate synthetic data for real-life projects. iblog.ridge-i.com/entry/2024/02/22/130409. Cited 16 March 2024 (2024)
74. Rombach, R., Blattmann, A., Lorenz, D., Esser, P., Ommer, B.: High-resolution image synthesis with latent diffusion models. In: Proceedings of the International Conference on Computer Vision and Pattern Recognition (CVPR), pp. 10684–10695 (2022)
75. Rothman, D.: Transformers for Natural Language Processing and Computer Vision, 3rd edn. Packt Publishing, Birmingham, UK (2024)
76. Saragih, D., Tyrrell, P.: Using diffusion models to generate synthetic labelled data for medical image segmentation. Int. J. Comput. Assisted Radiol. **19**, 1615–1625 (2024). arxiv.org/abs/2310.16794. Cited 25 March 2024
77. Shorten, C., Khoshgoftaar, T.M.: A survey on image data augmentation for deep learning. J. Big Data (2019). journalofbigdata.springeropen.com/articles/10.1186/s40537-019-0197-0. Cited 11 March 2024
78. Sohl-Dickstein, J., Weiss, E.A., Maheswaranathan, N., Ganguli, S.: Deep unsupervised learning using nonequilibrium thermodynamics. In: Proceedings of the International Conference on Machine Learning, vol. 37, no. C (2015). proceedings.mlr.press/v37/sohl-dickstein15.pdf. arxiv.org/abs/1503.03585. Cited 11 March 2024
79. Strang, G.: Introduction to Linear Algebra, 5th edn. Wellesley-Cambridge Press, Wellesley, MA, USA (2016). math.mit.edu/~gs/linearalgebra/ila5/indexila5.html. Cited 13 March 2024
80. Taylor, L., Nitschke, G.: Improving deep learning with generic data augmentation. In: IEEE Symposium on Series on Computational Intelligence, pp. 1542–1547 (2019)
81. Taylor, P.: Amount of data created, consumed, and stored 2010–2020, with forecasts to 2025 (2023). www.statista.com/statistics/871513/worldwide-data-created. Cited 3 March 2024
82. Trask, A.: Grokking Deep Learning. Manning, Shelter Island, NY, USA (2019)
83. Voetman, R., Aghaei, M., Dijkstra, K.: The big data myth: using diffusion models for dataset generation to train deep detection models (2023). arxiv.org/abs/2306.09762. Cited 11 March 2024
84. Wan, Z., et al.: Efficient large language models: a survey. Trans. Mach. Learn. Res. (TMLR) (2023). arxiv.org/abs/2312.03863. Cited 21 July 2024
85. Wang, D., Cui, X., Salcudean, S., Wang, Z.J.: Generalizable neural radiance fields for novel view synthesis with transformer (2022). arxiv.org/abs/2206.05375. Cited 26 March 2024
86. Wang, S., Minku, L.L., Yao, X.: Dealing with multiple classes in online class imbalance learning. In: Proceedings of the International Joint Conference on Artificial Intelligence (2016). www.cs.bham.ac.uk/~wangsu/documents/papers/IJCAI16.pdf. Cited 8 July 2024
87. Watkins, D.: Fundamentals of Matrix Computations, 3rd edn. Wiley, New York, USA (2010)
88. Wen,Q., Sun, L., Yang, F., Song, X., Gao, J., Wang, X., Xu, H.: Time series data augmentation for deep learning: a survey. In: Proceedings of the International Joint Conference on Artificial Intelligence, Survey Track, pp. 4653–4660 (2021). www.ijcai.org/proceedings/2021/631. Cited 11 March 2024
89. Werness, B., Wu, R.: Appendix: mathematics for deep learning (Chapter 22). In: Zhang, A., Lipton, Z. C., Li, M., Smola, A.J. (eds.) Dive into Deep Learning. Cambridge University Press, Cambridge, UK (2023). d2l.ai/chapter_appendix-mathematics-for-deep-learning/index.html. Cited 13 March 2024
90. Yang, H., Hong, L., Li, A., Hu, T., Li, Z., Lee, G.H., Wang, L. ContraNeRF: generalizable neural radiance fields for synthetic-to-real novel view synthesis via contrastive learning (2024). arxiv.org/abs/2303.11052. Cited 26 March 2024
91. Yang, S., Xiao, W., Zhang, M., Guo, S., Zhao, J. Shen, F.: Image data augmentation for deep learning: a survey (2023). arxiv.org/abs/2204.08610. Cited 11 March 2024
92. Zhang, J., Shi, P., Gu, Z., Zhou, Y., Wang, Z.: Semantic-human: neural rendering of humans from monocular video with human parsing (2023). arxiv.org/abs/2308.09894. Cited 26 March 2024
93. Zhao, W.X., et al.: A survey of large language models (2023). arxiv.org/abs//2303.18223. Cited 20 July 2024

94. Zhong, Z., Zheng, L., Kang, G., Li, S., Yang Y.: Random erasing data augmentation. In: Proceedings of the Associate for the Advancement of Artificial Intelligence (AAAI) Conference on Artificial Intelligence (2020). ojs.aaai.org/index.php/AAAI/article/view/7000. Cited 11 March 2024
95. Zhou, Y., Guo, C., Wang, X., Chang, Y., Wu, Y.: A survey on data augmentation in large model era (2024). arxiv.org/abs/2401.15422. Cited 10 July 2024

Chapter 2
Multiparty Computation

Abstract This chapter introduces the concept of multiparty computation and discusses some simple examples that serve as inspirations for the development of federated learning. The idea of multiparty computation is to have many parties, each of whom has only a small data set, agree to pool together their data resources to conduct analysis that would otherwise be impossible or lead to results having no practical value. In the most basic multiparty computation framework, each participating party sends encrypted, raw data to an external trusted authority for data aggregation, management, and analysis. In some circumstances, all parties may be unable to agree on a suitable external authority. In view of this difficulty, some multiparty computation frameworks without a trusted authority have been developed, but they involve generation and processing of more data and more computational work. These multiparty computation frameworks are technically viable solutions for small data sets. However, for applications involving training of AI models, they would involve encryption, transmission, processing, and decryption of such large volumes of data and would be riddled with so many legal business and technical challenges that real-world deployment would be impossible.

2.1 Introduction

The previous chapter discussed how institutions incur substantial costs in creating and storing their data. To recoup these costs, they would like to use data analytics and AI to improve their operations. Unfortunately, a major bottleneck for many institutions is the lack of high-quality data for training AI models. Data augmentation using replication or generating synthetic data may be possible if there is sufficient data with relevant information to serve as a template for producing new data. However, alternative approaches must be used to address inadequate data volume or extremely imbalanced data sets.

M. Kobayashi, *Federated Learning*, ICIAM2023 Springer Series 4,
https://doi.org/10.1007/978-981-96-9223-1_2

This chapter introduces *secure multiparty computation*[1] [5, 9, 10] (or *multiparty computation*), a framework that may help institutions resolve issues with data paucity. The framework can be used when many different parties with small volumes of high-quality data of the same format come to an agreement to pool together their resources to conduct data analytics for a specific purpose. Multiparty computation is based on the belief that "*nothing beats real data*", and provides a means for multiple parties "*to share data for computing tasks without revealing each other's data. All parties are privy to the output of the computing tasks, but no party learns anything about others*".[2]

2.2 Multiparty Computation with a Trusted Authority

The simplest type of multiparty computation framework (shown in Fig. 2.1) relies on a trusted external authority, agreed upon by all participating parties. It consists of the following steps:

- Each participating party
  - encrypts its own raw data and
  - sends the encrypted data to an external authority.
- The external authority
  - decrypts the data received from each participating party;
  - processes the data to be in the same format; and
  - uses the aggregated data to train an AI model.
- Results from analysis using the global data set are shared with all participating parties.
- Raw data from individual parties is *never* shared with other parties.

This simple multiparty computation framework is very difficult to deploy, in practice, in scenarios that involve joint training of an AI model using data aggregated from several sources. The massive volumes of data that must be encrypted and decrypted require considerable computational work. In general, more secure methods require more computation. The speed of current computers is too slow to enable very high levels of secure encryption.

A second difficulty is the computational expense for pre-processing massive volumes of data for training an AI model. Different institutions collect and store their

[1] Wikipedia page on *secure multiparty computation*: en.wikipedia.org/wiki/Secure_multiparty_computation. Cited 11 Jan 2024.

[2] IEEE Digital Privacy Special Interest Group (SIG) website: What is multiparty computation?: digitalprivacy.ieee.org/publications/topics/what-is-multiparty-computation. Cited January 11, 2024.

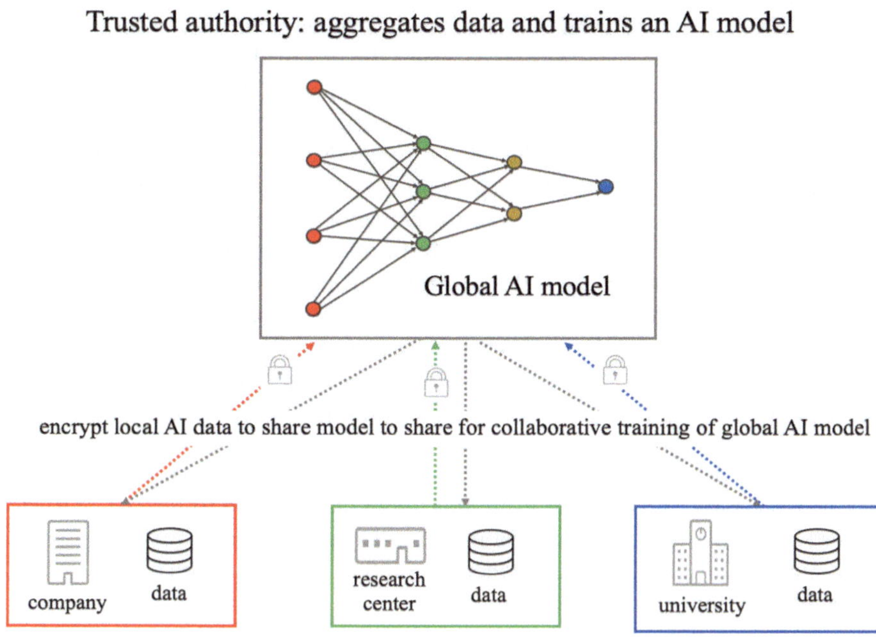

Fig. 2.1 Example of a simple multiparty computation framework for collaboratively training an AI model. Each of the three institutions (a company, research center, and university) has insufficient data for training an AI model. The institutions agree to encrypt and send their data to a trusted external authority. The trusted authority receives data from individual institutions; aggregates and formats the data; trains an AI model using the aggregated data; and sends (or gives access to) the trained AI model to each participating institution. The trusted authority *never* shares the raw data with others

data in different formats, so considerable pre-processing work may be needed to convert data from multiple sources to the same format. The cost of this data processing may be prohibitive for many institutions and businesses.

A third difficulty is updating the AI model in a timely manner. Additional sets of recent data must be encrypted and sent to the external authority. The external authority must decrypt and reformat the data to train an updated AI model. Updating a model, i.e., retraining, will be another long and costly computation. Some or all of the earlier (original training) data must be securely stored for retraining an updated model, adding to storage costs and data security concerns.

A fourth difficulty in deploying simple multiparty computation is the rise of data theft,[3] particularly with the cost per breach (monetary and reputation) surging for institutions with large data sets.[4] The invention of new types of theft, such as

[3] Statista, Annual number of data compromises and individuals impacted in the United States from 2005 to 2023 (February 12, 2024): www.statista.com/statistics/273550/data-breaches-recorded-in-the-united-states-by-number-of-breaches-and-records-exposed/.

[4] Cost of a Data Breach Report 2023, IBM Corporation: www.ibm.com/security/data-breach.

ransomware, has become an additional concern. Data theft may occur during storage or transmission and is a concern even when the stolen data is encrypted. Many encryption methods currently in use are not quantum-safe.[5] The forthcoming advent of quantum computers and their use by hostile parties is expected to become reality in the not-so-distant future.[6,7] The United States National Institute of Standards and Technology (NIST)[8] issued a *Report on Post-Quantum Cryptography* [3] that predicts that the first breaches to modern encryption standards, such as 2048-bit cryptography, may occur as early as 2030. In other words, encrypted data stolen today may be decrypted easily in the near future using quantum computers.

A fifth difficulty is bringing all parties together to agree on a capable external authority that is willing to deal with challenges associated with the cost and logistics of receiving, processing, storing and discarding the data. Many suitable institutions may be capable, but unwilling to serve as the authority due to the difficult and somewhat thankless nature of the job. For some highly select, critical applications (e.g., financial fraud detection, national security, medical research), government agencies may be reasonable, trustworthy institutions with an incentive to serve as a trusted authority to fulfill their social obligation to tax-paying citizens.

To summarize, real-world deployment of multiparty computation frameworks can be challenging for several reasons, such as

- the massive volume of data that must be encrypted/decrypted and transmitted;
- the massive computational work to pre-process data from multiple sources before training an AI model;
- the high computation and storage costs of keeping AI models up to date;
- the need for agreement by all parties on a capable external authority to receive, store, secure, format, and analyze data from multiple parties;
- the need for methods to reduce the risk of theft during transmission and storage of encrypted data; and
- the need to comply with government and corporate privacy regulations regarding personal data.

[5] Quantum-safe refers to cryptography that "*secures sensitive data, access, and communications for the era of quantum computing*". From: https://www.ibm.com/topics/quantum-safe-cryptography.

[6] openquantumsafe.org.

[7] Forbes Technology Council post: How quantum computing will transform cybersecurity (January 4, 2021): www.forbes.com/sites/forbestechcouncil/2021/01/04/how-quantum-computing-will-transform-cybersecurity/?sh=2b97253a7d3f.

[8] nist.gov.

2.3 Multiparty Computation Without a Trusted Authority

In some circumstances, it is difficult for all parties to agree upon a suitable external trusted authority. In view of this common impasse, extensions of the basic multiparty computation framework that eliminate the need for an external authority have been developed [1, 2].

A famous and simple textbook example (shown in Fig. 2.2) is a framework for three people (Alice, Bob, and Eve) who would like to compute their average salary without revealing the salary of any individual to others [9]. Alice chooses a very large random number and adds it to her salary. Alice sends the sum to Bob, who adds his salary and sends the sum to Eve. Eve adds her salary and sends the sum to Alice, who subtracts the very large random number and divides by three to find the average salary. Alice shares the results with Bob and Eve. Note that this framework assumes that all parties are honest, the number of participants is known, and everyone must agree that one person (in this case, Alice) will select the random number, then start and end the computations.

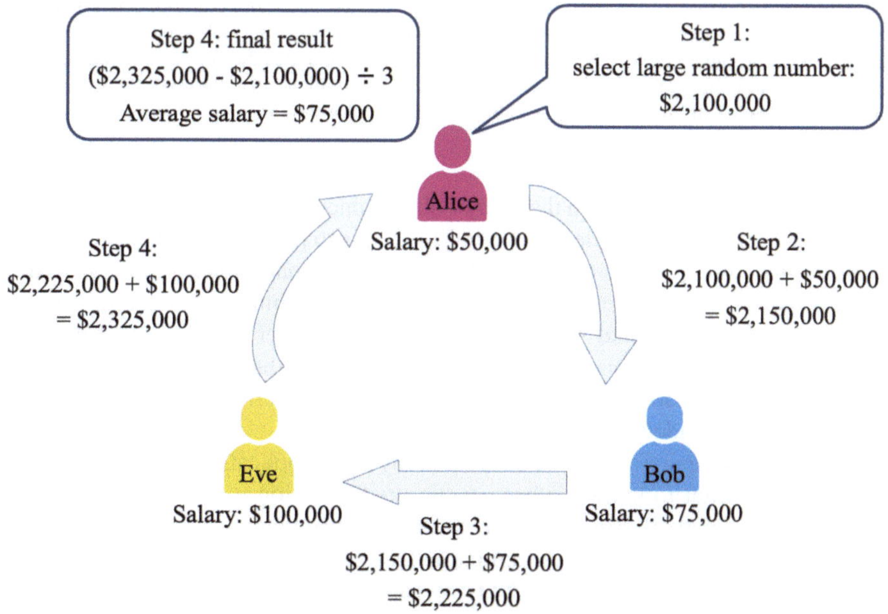

Fig. 2.2 Textbook example of a simple multiparty computation framework with no trusted external authority. Alice, Bob, and Eve would like to know their average salary, but none of them wants to reveal his/her own salary to others. Alice chooses a very large random number and adds it to her salary. Alice sends the sum to Bob, who adds his salary and sends the sum to Eve. Eve adds her salary and sends the sum to Alice, who subtracts the very large random number and divides by three to find the average salary. Alice shares the results with Bob and Eve. This framework assumes all parties are honest

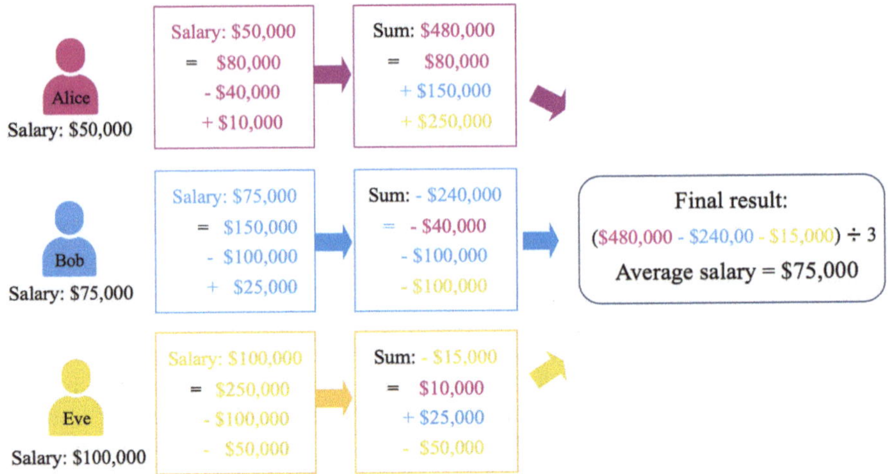

Fig. 2.3 A second textbook example of a simple multiparty computation framework with no trusted external authority. Alice, Bob, and Eve would like to know their average salary, but none of them wants to reveal his/her own salary to others. Each person selects three numbers that add up to his/her own salary. Then each participant sends the first of the three numbers to one person and the second of three numbers to the other person. The participant retains and stores the third number. After all numbers have been distributed, everyone computes the sum of the numbers they received with the one they retained and shares the numbers with others. The average salary is the sum of everyone's sum divided by three. This framework assumes all parties are honest

If everyone cannot agree on the idea of having one person having so much power, an alternate textbook example (shown in Fig. 2.3) is to have each person select three numbers that add up to his/her own salary. Then each participant sends the first of the three numbers to one person and the second of three numbers to the other person. The participant retains and stores the third number. After all numbers have been distributed, everyone computes the sum of the numbers they received with the one they retained and shares the numbers with others. The average salary is the sum of everyone's sum divided by three.

Multiparty computation can be used for many other applications [9]. For example, someone may want to know his/her genetic predisposition to cancer. A multiparty computation framework can securely compare the individual's DNA with a database of DNA of cancer patients to determine if there are unusually close matches and return only the type of cancer and NOT reveal any other genetic information about the individual. Another example is a trading platform, which keeps track of bids and offers. The platform notifies buyers and sellers when a match has been made and the price of the transaction. However, the platform never reveals the values of bids and offers to anyone since the information could be exploited to artificially raise prices by the seller or lower the payment offered by the buyer.

As in the examples above, frameworks without an external trusted authority generally increase computational requirements. In addition, the examples in this chapter assume that all parties are honest. It is beyond the scope of this monograph to delve deeply into the history and development of more sophisticated multiparty computational frameworks that can accommodate one or more dishonest participants. Interested readers are referred to publications and textbooks on the topic, such as Choudhury and Patra [4], Cramer and Damgård [5], Cramer, Damgård and Nielsen [6], Evans, Kolesnikov and Rosulek [7], Laud and Kamm [8], Zhao et al. [11], and Zhou at al. [12]. In general, algorithms with a high level of security that can also take into account possible rogue participants tend to be technically complex, computationally expensive, and impractical to implement for large data sets, such as those used to train AI models.

A final note is that in many real-world scenarios multiparty computation cannot be deployed when the raw data includes personal information about customers or employees. Many countries prohibit institutions and businesses from sharing raw data containing personal information with external parties.[9,10,11] Waiting for changes in laws to accommodate new frameworks is not a viable strategy. The worldwide trend is an *increase* (rather than decrease) in laws to restrict practices in data storage, transmission, and sharing.

References

1. Bogetoft, P., Christensen, D.L., Damgård, I., Geisler, M., et al.: Secure multiparty computation goes live. In: Proceedings of the International Conference on Financial Cryptography and Data Security, Lecture Notes in Computer Science, vol. 5628, 325–343. Springer, Berlin, Germany (2009)
2. Breuer, M., Meyer, U., Wetzel, S.: Introducing a framework to enable anonymous secure multiparty computation in practice. In: Proceedings of the International Conference on Privacy, Security and Trust (2021). arxiv.org/abs/2111.02269. Cited 18 March 2024
3. Chen, L., Jordan, S., Liu, Y.-K., Moody, D., et al.: Report on post-quantum cryptography. NISTIR 8105, National Institute of Standards and Technology, Gaithersburg, MD, USA (2016). nvlpubs.nist.gov/nistpubs/ir/2016/nist.ir.8105.pdf. Cited 5 April 2024
4. Choudhury, A., Patra, A.: Secure Multi-party Computation Against Passive Adversaries. Springer Nature, Berlin, Germany (2022)
5. Cramer, R., Damgård, I. B.: Multiparty computation, an introduction (2004). www.researchgate.net/publication/226390713_Multiparty_Computation_an_Introduction. Cited 11 May 2024
6. Cramer, R., Damgård, I.B., Nielsen, J.B.: Secure Multiparty Computation and Secret Sharing. Cambridge University Press, Cambridge, UK (2015)
7. Evans, D., Kolesnikov, V., Rosulek, M.: A Pragmatic Introduction to Secure Multi-party Computation. Now Publishers, Hanover, MD, USA (2019)
8. Laud, P., Kamm, L.: Applications of Secure Multiparty Computation. IOS Press, The Netherlands, Amsterdam (2015)

[9] Protection of Personal Information Act, Japan: www.cas.go.jp/jp/seisaku/hourei/data/APPI.pdf.

[10] Personal Information Protection Commission Japan, Laws and policies: www.ppc.go.jp/en/legal/.

[11] General Data Protection Regulation (GDPR) of the European Union: gdpr.eu.

9. Lindell, Y.: Secure multiparty computation. Commun. ACM **64**(1), 86–96 (2020). https://doi.org/10.1145/3387108. Cited 3 April 2024
10. Lindell, Y., Pinkus, B.: Secure multiparty computation for privacy-preserving data mining. J. Priv. Confid. **1**(1), 197 (2009). journalprivacyconfidentiality.org/index.php/jpc/article/view/566. Cited 11 May 2024
11. Zhao, C., Zhao, S., Zhao, M., Chen, Z., et al.: Secure multi-party computation: theory, practice and applications. Inf. Sci. **476**, 357–372 (2019)
12. Zhou, I., Tofigh, F., Piccardi, M., Abolhasan, M., et al.: Secure multi-party computation for machine learning: a survey. IEEE Access **12**, 53881–53899 (2024)

Chapter 3
Edge Computing

Abstract *Edge computing* is a distributed computing model in which data processing takes place on local devices, closer to where it is being generated, reducing latency and speeding up data processing. Advances in hardware and economies of scale have made it possible for edge computing devices to be embedded in everyday consumer products to process large volumes of data quickly and produce results in near real time. Edge computing enables service providers to analyze customer preferences, behaviors, etc., without breaching personal privacy regulations. The ostensible reason that service providers gather troves of consumer data is to better understand customers and to improve customer satisfaction and safety. The information may also provide insights for new and enhanced services and products. As edge computing devices have proliferated, at least four important categories of the framework have emerged: edge computing on mobile devices, fog computing, cloudlets, and mobile edge computing networks. Definitions of these categories, their differences and individual strengths will be discussed. Examples of products and services suited to each category will also be given.

3.1 Distributed Computing

The origins of edge computing date back to *distributed computing*[1,2,3,4] models that were developed in the 1960s, when computers had only modest capabilities and scalability was needed to solve complex scientific problems. The idea of distributed computing is to have multiple computers work together to act as if they were a single, more powerful computer. Since its inception, distributed computing has become

[1] Amazon Web Services on *distributed computing*: aws.amazon.com/what-is/distributed-computing.

[2] Open Cirrus, "What is Distributed Computing, its Pros and Cons?": opencirrus.org/distributed-computing-pros-cons.

[3] Prof. Eric Roberts, Stanford Univ.: cs.stanford.edu/people/eroberts/courses/soco/projects/distributed-computing.

[4] Wikipedia page on *Distributed Computing*: en.wikipedia.org/wiki/Distributed_computing.

M. Kobayashi, *Federated Learning*, ICIAM2023 Springer Series 4,
https://doi.org/10.1007/978-981-96-9223-1_3

commonplace in a wide range of applications, such as healthcare and life sciences, 3D modeling, visualizations of atoms and molecules, computational fluid dynamics, computer-aided engineering, and simulations of financial portfolio risks [11, 15, 24, 29, 30, 32, 44, 47–49].

Although parallel computing and distributed computing share some common features, they have some important differences. Most notably, in parallel computing processors have access to shared memory for information exchange, while in distributed computing each processor has its own, private memory (as illustrated in Fig. 3.1).

Although work is required to synchronize computations, distributed computing has some distinct advantages over parallel computing, for example:

- Distributed computing can accommodate much larger storage and memory, faster compute, and higher bandwidth than a single machine.
- It is easy to scale a distributed system up or down, depending on computational requirements.
- When a parallel machine fails, the entire system goes down. However, failure in a machine in a distributed system will only result in a shutdown of the problematic machine, and work can continue on other machines. The system can be set up so that work assigned to the failed machine will be rerouted to the remaining working machines. In short, distributed systems tend to have higher fault tolerance.
- Distributed systems tend to be less expensive since a cluster of inexpensive machines is usually less costly than a single, parallel processing machine.

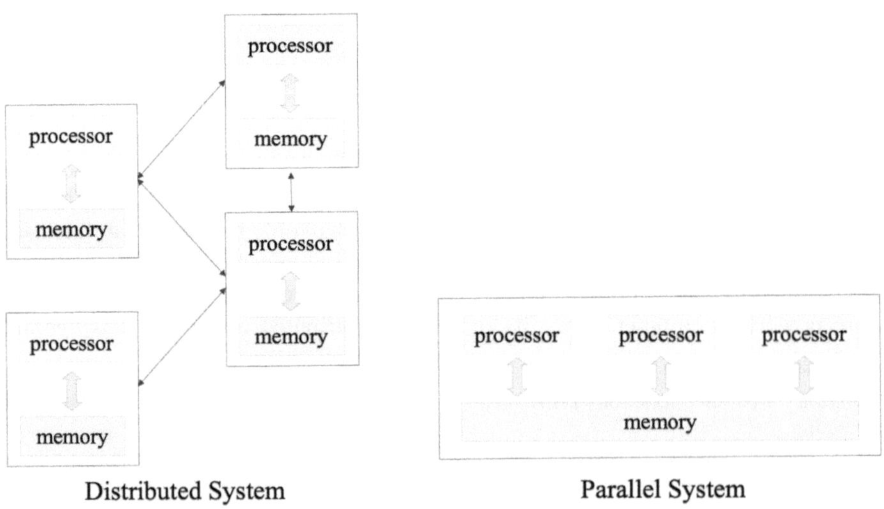

Fig. 3.1 Distributed computing versus parallel computing. In parallel computing processors have access to shared memory for information exchange, while in distributed computing each processor has its own, private memory

Sometimes distributed computing is not suited for an application. Parallel computing has some distinct advantages over distributed computing, for example:

- Parallel computing tends to be better for large-scale scientific simulations since it can carry out the computations efficiently, with less communication overhead and a clear point of central control. Distributed computing tends to have a higher communication overhead and is less efficient for scientific computing.
- Parallel computing systems tend to be easier to deploy, maintain, and troubleshoot than distributed computing systems. However, quite often, engineers have to learn a language specifically developed for a parallel computer.
- Security is easier to manage in parallel computing systems. Usually special software is required for handling security and communications in distributed systems.

Whether a parallel or distributed computing system is better suited for use depends on the problem to be solved, data pipeline, and application scenario. Listing and prioritization of the important points (cost, speed, human labor, scalability, security requirements, etc.) in deployment and maintenance will help users and owners of the system make a well-informed decision.

3.2 Edge Computing

Edge computing is a distributed computing model in which data processing takes place on devices at the "edge of a network", closer to where it is being generated, reducing latency and speeding up many applications[5,6,7] [3, 5–8, 16, 23, 25, 28, 38, 42, 46]. Although the underlying ideas behind edge computing existed decades ago, the value of the paradigm came to fruition with its deployment alongside the proliferation of connected consumer devices with inexpensive and powerful computing capabilities [14, 26, 27, 36, 39–41, 51]. Examples of these devices include smart connected vehicles, cell phones, smart watches, smart glasses, smart VR headsets, smart automobile services, and smart homes with smart meters (as shown in Fig. 3.2).

These days, many companies use edge computing to analyze consumer preferences. Figure 3.2 gives an overview of a typical edge computing framework, with cell phones serving as an example. Only summaries of the analyses are sent to the cloud, where companies aggregate the information to understand overall consumer behaviors. Edge computing has several advantages over traditional methods which normally transmit user data to a central server on the cloud. The framework

[5] Cloudwards website on edge computing: www.cloudwards.net/what-is-edge-computing.

[6] IBM website on edge computing: www.ibm.com/topics/edge-computing.

[7] Microsoft website on edge computing: azure.microsoft.com/en-us/resources/cloud-computing-dictionary/what-is-edge-computing.

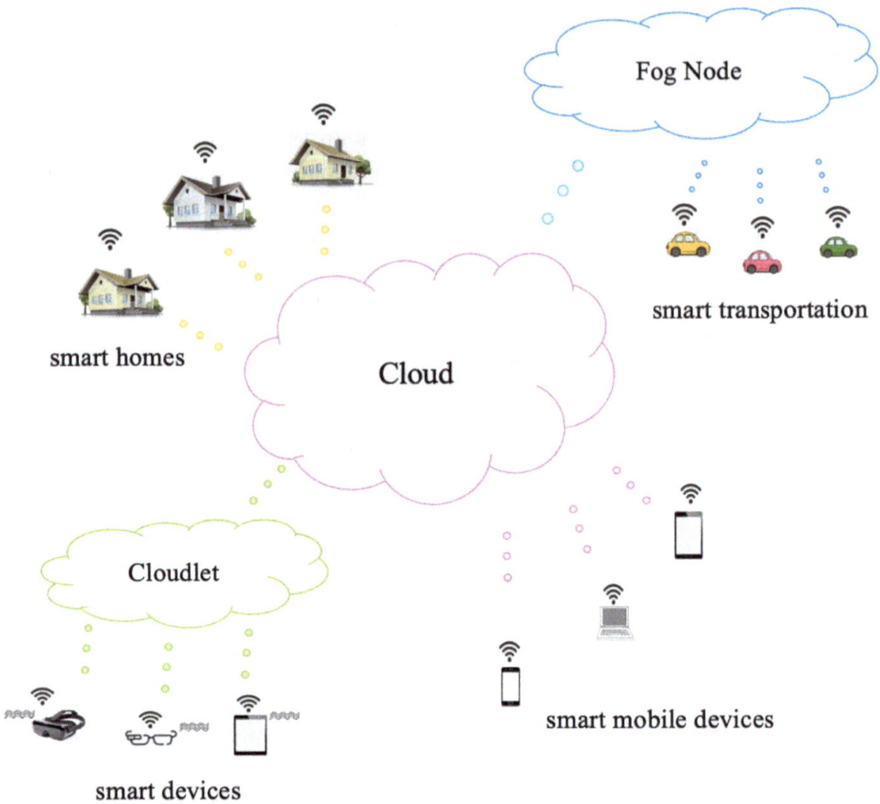

Fig. 3.2 Edge computing. Edge computing is a distributed computing model in which data processing takes place on devices on the "edge of a network", closer to where it is being generated, reducing latency and speeding up many applications. Some examples of common, everyday connected devices using edge computing are smart/connected vehicles, cell phones, mobile devices, smart glasses, and smart VR headsets

- offers higher protection of privacy of personal data since raw data remains on the consumer's own, personal device; only results from analysis of the data are sent to the service provider, usually in encrypted form.
- reduces latency since the data is processed close to where it is created.
- reduces transmission, storage, security, data processing, and computational costs for the service provider since only data summaries are sent from edge devices. The summaries are typically orders of magnitude smaller than the raw data.

In short, edge computing enables faster, less expensive and more secure analysis of data from multiple sources, which may number in the millions or billions.

As deployments of edge computing systems have proliferated, at least four important categories of the framework have emerged [17, 45]:

- edge computing on mobile devices,
- fog computing,
- cloudlets, and
- mobile edge computing networks.

They are described in the ensuing subsections.

3.2.1 Edge Computing on Mobile Devices

Edge computing on mobile devices is processing and analysis on the mobile device(s) on which the data is generated, e.g., on cell phones or tablets (as shown in Fig. 3.3) [13, 17]. Although computing capabilities on mobile devices have increased significantly in recent years, they are still not powerful enough for large-scale computations.

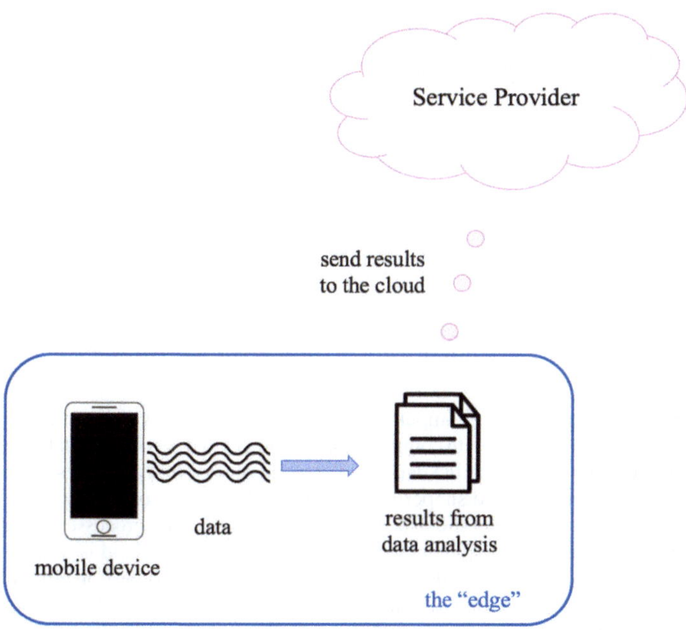

Fig. 3.3 Edge computing by mobile devices. A mobile device generates data. The data is analyzed on the device, and only the results from analysis (not the raw data) are sent to the service provider in the cloud

3.2.2 Fog Computing

The *fog computing*[8] framework enhances edge computing on devices by adding a computing infrastructure with computational resources and services for data pre-processing between devices that are generating the data and the cloud service that processes and analyzes the data (as shown in Fig. 3.4) [37]. When the distance between the data being generated and the cloud is far, and the latency is unacceptably high, fog nodes can provide support to speed up the data ingestion pipeline. Since fog nodes lie very close to the source of the new data, they can quickly pre-process the data before sending it to the cloud for heavier computing tasks and storage.

One of the main drawbacks of fog computing is security. It is susceptible to IP address spoofing and man-in-the-middle attacks. Since data is processed outside of the owner's device, privacy issues and authentication concerns also come into play. Security of the wireless network used for transmitting data is yet another concern.

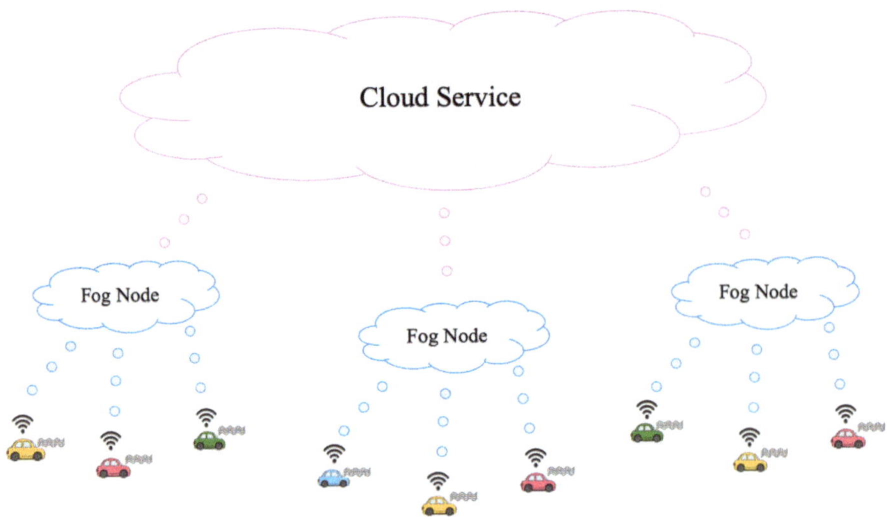

Fig. 3.4 Fog computing. The fog computing framework enhances edge computing on devices by adding a computing infrastructure with computational resources and services for data pre-processing between devices that are generating the data and the cloud service that processes and analyzes the data. For example, smart cars need support from additional computational infrastructure between devices in them that are generating the data and the cloud service that processes and analyzes the data. The distance between the smart cars generating the data and the cloud is far, and the latency is unacceptably high. Fog nodes provide support to speed up the data ingestion pipeline. Since fog nodes lie very close to the source of new data, they can quickly pre-process the data before sending it to the cloud for heavier computing tasks and storage. The example in this figure shows fog nodes enhancing the safety, ease of driving, and entertainment for occupants of smart vehicles

[8] According to Cisco, the term "fog computing" was invented by the company in 2014 to describe "the decentralization of computing infrastructure, or bringing the cloud to the ground": www.cisco.com/c/en/us/solutions/computing/what-is-edge-computing.html.

A third drawback is the physical location boundaries that support fog networks, in contrast to the "anytime, anywhere" access paradigm of cloud computing [12].

The *OpenFog Consortium*[9] is a consortium of high-tech companies and academic institutions that promotes the advancement and dissemination of fog computing and the standardization of architectures.[10] Since its establishment in 2015, the Consortium has grown from six members (Cisco Systems, Intel, Microsoft, Princeton University, Dell, and ARM Holdings) to 57 members in North America, Asia, and Europe.

One of the earlier, commercially deployed scenarios of fog computing was for the enhancement of safety, ease of driving, and entertainment for occupants of smart vehicles.[11] Recently, connected vehicles are incorporating increasingly sophisticated features that are computationally demanding. At the same time, expectations of consumers are high regarding quality of experience (QoE). Changes in the edge computing paradigm for connected vehicles are rapidly being developed ([17], Chap. 5). Drivers of commercial vehicles (e.g., delivery trucks, long-haul vans, buses, taxis, ambulances) will have different and more demanding needs than ordinary consumers to improve efficiency and safety while at work.

Other applications of fog computing include healthcare (patient monitoring), energy management (smart grid management of alternative energy sources), agriculture (smart crop and irrigation system management), and smart home management (security and fire alarm systems).

3.2.3 Cloudlets

Cloudlets are mini, cloud-like data centers with powerful computing resources located at the "edge" of networks. Their main purpose is to support resource-intensive mobile applications (such as augmented reality systems, smart glasses, intelligent virtual assistants) that require fast response times [9, 31]. The term "cloudlet" was coined by Satyanarayanan, Bahl, Cáceres and Davies [43] and first implemented as a project, *Elijah*, at Carnegie Mellon University.[12] Cloudlets are region-specific, meaning that when a mobile device moves from a region serviced by one cloudlet to a region serviced by a different cloudlet, the former cloudlet must hand off the user's virtual machine to the latter cloudlet (as shown in Fig. 3.5).

---

[9] opcfoundation.org.

[10] OpenFog Reference Architecture for Fog Computing: www.iiconsortium.org/pdf/OpenFog_Reference_Architecture_2_09_17.pdf.

[11] geeksforgeeks.org, "Difference Between Edge Computing and Fog Computing" (2002): www.geeksforgeeks.org/difference-between-edge-computing-and-fog-computing [4, 17].

[12] elijah.cs.cmu.edu.

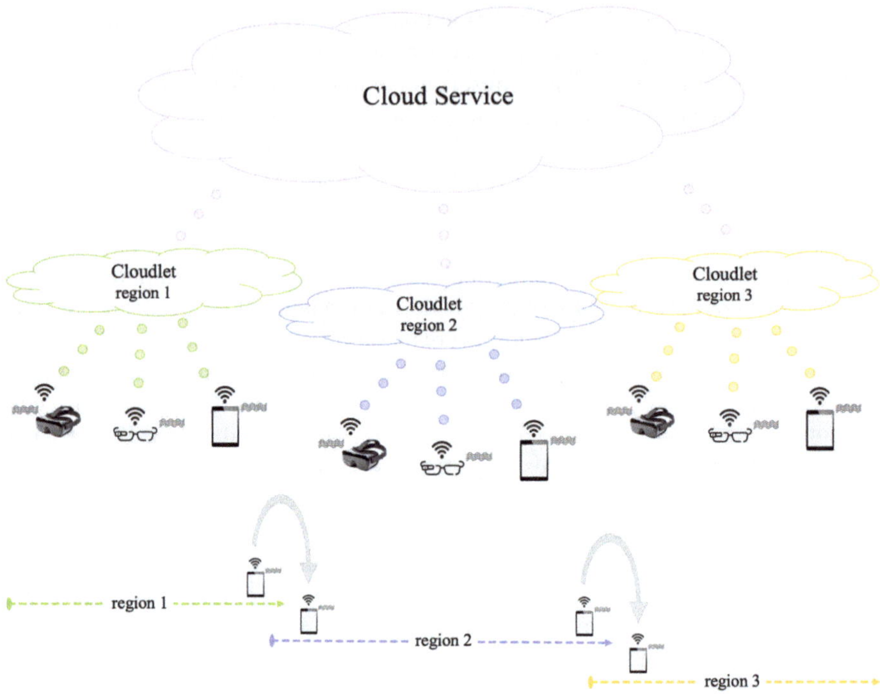

Fig. 3.5 Cloudlets. *Cloudlets* are mini cloud-like data centers with powerful computing resources located at the "edge" of networks. Their main purpose is to support resource-intensive mobile applications (such as augmented reality systems, smart glasses, intelligent virtual assistants) that require fast response times. Cloudlets are region-specific, so that when a mobile device moves from a region serviced by one cloudlet to a region serviced by a different cloudlet, the former cloudlet must hand off the user's virtual machine to the latter cloudlet

In addition to reducing the workload sent to the cloud, the cloudlet framework has other attractive features, for example:

- Cloudlets are highly scalable. When the number of customers or IoT devices increases in a region, more cloudlets can be quickly added to cover the computing needs for that region.
- Cloudlets can also provide more user privacy than cloud computing models since the number of users for each cloudlet is limited, and data is stored locally, in cloudlets, rather than in the cloud.
- Finally, cloudlets are more effective for controlling the scope of outages. A catastrophic failure in the cloud can wipe out services to millions of users, while a catastrophic failure of a cloudlet does not lead to denial of services to users. When a cloudlet fails, users are redirected to another nearby cloudlet.

Famous examples of mobile applications supported by cloudlets include augmented reality systems, smart glasses, intelligent virtual assistants, and cloud games.[13]

Sometimes, the three categories of edge computing (edge computing on mobile devices, fog computing, and cloudlets) overlap, and distinctions between them can be a bit ambiguous. Some applications use edge computing from only one category, while others may rely on edge computing from more than one. Fog computing is primarily used for pre-processing data very close to where it is being generated before sending it to the cloud for analysis. In general, fog nodes do not have to carry out heavy computational work, like cloudlets. Cloudlets are like mini-data centers, where computationally demanding simulations and analysis can be performed. Cloudlets are particularly useful for supporting resource-intensive mobile applications that need near-real-time responses, like augmented reality systems and gaming.

Hackers have begun to attack edge computing systems as their prevalence has increased. Some reviews of studies on security of edge computing are [21, 22, 50].

3.2.4 Mobile Edge Computing Networks

A variety of edge computing paradigms have been developed and deployed, but with constantly evolving needs, interesting and useful new ideas are always being developed. A notable example is the deployment of mobile edge computing networks assisted by unmanned aerial vehicles, such as drones [1, 2, 10, 17–20, 33–35, 52, 53]. Mobile networks can play a variety of roles, such as providing temporary computational support, forwarding larger computational work to terrestrial mobile servers, and forwarding computational work to established terrestrial servers.

These mobile networks are valuable tools in scenarios, including support of temporary hotspots with sudden surges in demand, communication during maintenance work in remote locations, and broken computing pipelines. Some real-world examples in which mobile edge computing networks have been deployed include the following:

- disaster relief,
- emergency rescue operations in remote areas,
- inspection of remote wind farms and off-shore wind turbine facilities,
- communication support in battlefields, and
- temporary support of edge computing support facilities that have broken down.

Some of the unique challenges with mobile edge computing networks are energy consumption minimization, completion time minimization, trajectory (flight path) optimization, and utility maximization (i.e., balancing the tradeoffs between computation energy consumption, time delay, and computational costs).

[13] Wikipedia page on cloudlets: https://en.wikipedia.org/wiki/Cloudlet.

References

1. Abbas, N., Zhang, Y., Taherkordi, A., Skeie, T.: Mobile edge computing: a survey. IEEE Internet Things J. **5**(1), 450–465 (2018). ieeexplore.ieee.org/document/8030322. Cited 4 September 2024
2. Abkenar, F.S., Ramezani, P., Iranmanesh, S., Murali, S., et al.: A survey on mobility of edge computing networks in IoT: state-of-the-art, architectures, and challenges. IEEE Commun. Surv. Tutor. **24**(4), 2329–2365 (2022)
3. Al-Dulaimy, A., Sharma, Y., Khan, M., Taheri, J.: Introduction to edge computing. Chapter 1 in [46] (2020). www.researchgate.net/publication/344218125_Introduction_to_edge_computing. Cited 11 May 2024
4. Behravan, K., Farzaneh, N., Jahanshahi, M., Seno, S.A.H.: A comprehensive survey on using fog computing in vehicular networks. Veh. Netw. **42**(18), 100604 (2023)
5. Bhopal, S.: Edge Computing: research and Outlook. Thesis, Dept. of Computer Science, California State Polytechnic University, Pomona, CA, USA, M.A (2020)
6. Cao, J., Zhang, Q., Shi, W.: Edge Computing: a Primer. Springer, Berlin, Germany (2018)
7. Cao, K., Liu, Y., Meng, G., Sun, Q.: An overview on edge computing research. IEEE Access **8**, 85714–85728 (2020)
8. Carvalho, G., Cabral, B., Pereira, V., Bernadino, J.: Edge computing: current trends, research challenges and future directions. Computing **103**, 993–1023 (2021)
9. Chen, J., Ran, X.: Deep learning with edge computing: a review. Proc. IEEE **107**(8), 1655–1674 (2019)
10. Cheng, N., Xu, W., Shi, W., Zhou, Y., et al.: Air-ground integrated mobile edge computing networks: architecture, challenges, and opportunities. IEEE Commun. Mag. **56**(8), 26–32 (2018)
11. Coulouris, G., Dollimore, J., Kindberg, T., Blair, G.: Distributed Systems: concepts and Design, 5th edn. Pearson Education, Boston, MA, USA (2012)
12. Dharshana, A.D., Ramya, K., Kanimozhi, N.: Fog computing: an overview. Int. J. Trend Res. Dev. **4**(1), 462–464 (2017). www.ijtrd.com/papers/IJTRD6615.pdf. Cited 4 September 2024
13. Dimou, A., Iliopoulos, C., Polytidou, E., Dhurandher, S.K., et al.: A comprehensive review on edge computing: Focusing on mobile users. In: Nicopolitidis, P., Misra, S., Yang, L.T., Zeigler, B., Ning, Z. (eds.) Advances in Computing, Informatics, Networking and Cybersecurity. Lecture Notes in Networks and Systems, vol. 289, pp. 121–152. Springer, Cham, Switzerland (2022)
14. França, R.P., Monteiro, A.C.B., Arthur, R., Iano, Y.: An overview of the edge computing in the modern digital age. In: Chang, W., Wu, J. (eds.) Fog/Edge Computing For Security, Privacy, and Applications. Advances in Information Security, vol. 83. Springer, Cham, Switzerland (2021)
15. Garg, V.K.: Elements of Distributed Computing. Wiley-IEEE Press, Hoboken, New Jersey, US (2022)
16. Gill, S.S., Golec, M., Hu, J., Xu, M., et al.: Edge AI: a taxonomy, systematic review and future directions (2024). arxiv.org/abs2407.04053v1. Cited 1 September 2024
17. Hsu, C.-H., Wang, S., Zhang, Y., Kobusińska, A.: Mobile Edge Computing. Simula Springer Briefs on Computing, Springer Nature, Switzerland (2022)
18. Hu, X., Wong, K., Yang, K., Zheng, Z.: UAV-assisted relaying and edge computing: scheduling and trajectory optimization. IEEE Trans. Wirel. Commun. **18**(10), 4738–4752 (2019)
19. Hua, M., Wang, Y., Wu, Q., Li, C., et al.: Energy optimization for cellular-connected multi-UAV mobile edge computing systems with multi-access schemes. J. Commun. Netw. **3**(4), 33–44 (2018)
20. Jeong, S., Simeone, O., Kang, J.: Mobile edge computing via a UAV-mounted cloudlet: optimization of bit allocation and path planning. IEEE Trans. Veh. Technol. **67**(3), 2049–2063 (2018)
21. Jin, X., Katsis, C., Sang, F., Sun, J., et al.: Edge security: challenges and issues (2022). arxiv.org/abs/2206.07164. Cited 20 May 2024

22. Kalariya, H., Patel, V., Shah, K.: A systematic literature review on edge computing security (2022). arxiv.org/abs/2212.04563. Cited 20 May 2024
23. Khan, W. Z., Ahmed, E., Hakak, S., Yaqoob, I., Ahmed, A.: Edge computing: a survey. Future Gener. Comput. Syst. **97**, 219–235 (2019). www.researchgate.net/publication/331362529_Edge_computing_A_survey. Cited 9 May 2024
24. Kleppmann, M.: Designing Data-Intensive Applications: the Big Ideas Behind Reliable, Scalable, and Maintainable Systems. O'Reilly Media, Sebastopol, CA, USA (2017)
25. Kollu, A., Chennam, K.K., Mahajan, D.: An empirical review on secure edge computing architecture. In: Kumar, A., Ghinea, G., Merugu, S (eds.) Proceedings of 2022 International Conference on Cognitive and Intelligent Computing (ICCIC), pp. 661–668. Springer, Berlin, Germany (2023)
26. Kong, L., Tan, J., Huang, J., Chen, G., et al.: Edge-computing-driven internet of things: a survey. ACM Comput. Surv. **55**(8), 1–41 (2022)
27. Kong, X., Wu, Y., Wang, H., Xia, F.: Edge computing for internet of everything: a survey. IEEE Internet Things J. **9**(23), 23472–23485 (2022)
28. Krishnasamy, E., Varrette, S., Mucciardi, M.: Edge computing: an overview of framework and applications. In: Report for Partnership for Advanced Computing in Europe (2020). prace-ri.eu/wp-content/uploads/Edge-Computing-An-Overview-of-Framework-and-Applications.pdf. Cited 9 May 2024
29. Kshemkalyani, A.D., Singhal, M.: Distributed Computing: principles, Algorithms, and Systems. Cambridge University Press, Cambridge, UK (2011)
30. Leopold, C.: Parallel and Distributed Computing: a Survey of Models. Paradigms and Approaches, Wiley-Interscience, Hoboken, NJ, USA (2000)
31. Liang, S., Jin, S., Chen, Y.: A review of edge computing technology and its applications in power systems. Energies **17**(13), 3230 (2024). https://doi.org/10.3390/en17133230. Cited 2 September 2024
32. Lynch, N.: Distributed Algorithms. Morgan Kaufmann, San Francisco, CA, USA (1997)
33. Lyu, J., Zeng, Y., Zhang, R.: UAV-aided offloading for cellular hotspot. IEEE Trans. Wirel. Commun. **17**(6), 3988–4001 (2018)
34. Mao, Y., You, C., Zhang, J., Huang, K., Letaief, K.B.: A survey on mobile edge computing: the communication perspective. IEEE Commun. Surv. Tutor. **19**(4), 2322–2358 (2017)
35. Motlagh, N.H., Bagaa, M., Taleb, T.: UAV-based IoT platform: a crowd surveillance use case. IEEE Commun. Mag. **55**(2), 128–134 (2017)
36. Nayak, S., Patgiri, R., Waikhom, L., Ahmed, A.: A review on edge analytics: issues, challenges, opportunities, promises, future directions, and applications. In: Digital Communications and Networks (2024). www.sciencedirect.com/science/article/pii/S2352864822002255. Cited 12 May 2024
37. Patel, S., Patel, R.: J. Eng. Sci. Technol. Rev. **15**(3), 63–83 (2022). www.jestr.org/downloads/Volume15Issue3/fulltext81532022.pdf. Cited 5 September 2024
38. Patra, B., Tamrakar, A., Sharma, R.: Edge computing: evolution, challenges, and future directions. Turkish J. Comput. Math. Educ. **10**(1), 741–745 (2019). turcomat.org/index.php/turkbilmat/article/view/14603. Cited 2 September 2024
39. Pérez, J., Díaz, J., Berrocal, J., López-Viana, R., González-Prieto, Á.: Edge computing: a grounded theory study. Computing **104**, 2711–2747 (2022). link.springer.com/article/10.1007/s00607-022-01104-2. Cited 12 May 2024
40. Risteska Stojkoska, B.L., Trivodaliev, K.V.: A review of internet of things for smart home: challenges and solutions. J. Clean. Prod. **140**(3), 1454–1464 (2017)
41. Rose, K., Eldridge, S., Chapin, L.: The internet of things: an overview. In: The Internet Society (2015). www.internetsociety.org/wp-content/uploads/2017/08/ISOC-IoT-Overview-20151221-en.pdf. Cited 12 May 2024
42. Safavat, S., Sapavath, N.N., Rawat, D.B.: Recent advances in mobile edge computing and content caching. Digit. Commun. Netwo. **6**(2), 189–194 (2020). www.sciencedirect.com/science/article/pii/S2352864819300227. Cited 2 September 2024

43. Satyanarayanan, M., Bahl, V., Cáceres, R., Davies, N.: The case for VM-based cloudlets in mobile computing. IEEE Pervasive Comput. **8**(4), 14–23 (2009)
44. Schlossnagle, T.: Scalable Internet Architectures. Sams Publishing, Carmel, IN, USA (2006)
45. Sheth, V.: Fog/Edge-computing and cloudlets. Medium.com (2019). medium.com.tharanignanasegaram/edge-computing-and-cloudlets-c4db6cc9b15a. Cited 18 April 2024
46. Taheri, J., Deng, S., (eds.).: Edge computing: models, Technologies and Applications. The Institution of Engineering and Technology (IET) Digital Library, London, UK (2020)
47. Tanenbaum, A.S., Van Steen, M.: Distributed Systems: principles and Paradigms, 2nd edn. Pearson, Prentice-Hall, Upper Saddle River, NJ, USA (2007)
48. Tel, G.: Introduction to Distributed Algorithms. Cambridge University Press, Cambridge, UK (2004)
49. Varela, C.A.: Programming Distributed Computing Systems: a Foundational Approach. MIT Press, Cambridge, MA, USA (2013)
50. Xiao, Y., Jia, Y., Liu, C., Cheng, X., et al.: Edge computing security: state of the art and challenges. Proc. IEEE **107**(8), 1608–1631 (2019). ieeexplore.ieee.org/documentl/8741060. Cited 20 May 2024
51. Yu, W., Liang, F., He, X., Hatcher, W., et al.: A survey on the edge computing for the internet of things. IEEE Access **6**, 6900–6919 (2017). ieeexplore.ieee.org/document/8123913. Cited 8 August 2024
52. Zhou, F., Hu, R., Li, Z., Wang, Y.: Mobile edge computing in unmanned aerial vehicle networks. IEEE Wirel. Commun. **27**(1), 140–146 (2020)
53. Zhou, F., Wu, Y., Hu, R., Qian, Y.: Computation rate maximization in UAV-enabled wireless powered mobile-edge computing systems. IEEE J. Sel. Areas Commun. **36**(9), 1927–1941 (2018)

Chapter 4
Federated Learning

Abstract *Federated learning* is a framework that enables multiple parties to collaboratively train AI models, while each party retains control of its own raw data, never sharing it with others. As such, it resolves some of the legal and technical challenges with basic multiparty computation. Federated learning relies on edge computing, a distributed computing model in which processing of raw data takes place on local devices, close to or where it was generated. Only results from local analysis using edge computing are sent to external data centers to collaboratively train a global AI model. Variations of the federated learning framework are used in a wide range of scenarios, such as *cross-device federated learning* on mobile devices by Internet companies to gain a better understanding of customer behaviors and preferences; *cross-silo federated learning* by different institutions with similar data that collaborate to increase the reliability of rare event identification (e.g., financial fraud, identification of diseases); *vertical federated learning* for aggregating information from different institutions with data about the same customers but different customer attributes, to gain a holistic understanding of customers; and *federated transfer learning* to increase the data sample and data attribute sizes by transferring knowledge across data sets.

4.1 Introduction

Federated learning is a relatively new type of multiparty computation that is increasingly receiving attention since the method resolves some legal, data transmission, and technical challenges with the simple, more traditional multiparty computation frameworks discussed earlier, in Chap. 2 [2, 3, 6, 17, 54, 56, 69–72, 92, 108]. Originally proposed, designed, and implemented independently by Chilimbi, Suzue, Apacible and Kalyanaraman [10] and Shokri and Shmatikov [79] in academic settings, federated learning is a framework for collaboratively training an AI model. It relies on edge computing, a distributed computing framework introduced in the previous chapter:

M. Kobayashi, *Federated Learning*, ICIAM2023 Springer Series 4,
https://doi.org/10.1007/978-981-96-9223-1_4

> (A federated learning) system enables multiple parties to jointly learn an accurate neural network model for a given objective without sharing their input data sets. (It) exploits the fact that the optimization algorithms used in modern deep learning, namely, those based on stochastic gradient descent, can be parallelized and executed asynchronously. (The) system lets participants train independently on their own data sets and selectively share small subsets of their models' key parameters during training. This offers an attractive point in the utility/privacy tradeoff space: participants preserve the privacy of their respective data while still benefitting from other participants' models and thus boosting their learning accuracy beyond what is achievable solely on their own inputs. – Shokri and Shmatikov [80]

Through the use of edge computing, federated learning ensures that raw data is processed close to where it was generated and is never shared with other participants during the collaborative training of an AI model. Only summary information during or after analysis (e.g., some model parameters during training, intermediate gradients during the training process, weights of local AI models) are encrypted and shared with external parties.

This chapter introduces three basic types of federated learning: horizontal (cross-device and cross-silo), vertical, and transfer learning. To simplify the descriptions of three different types of federated learning, we consider the scenario wherein each participating institution has data about its customers. The appropriate type of federated learning depends on the data and its attributes (as shown in Fig. 4.1):

- *horizontal federated learning*: Participating institutions have data on different customers. The data consists of the same customer attributes.
- *vertical federated learning*: Participating institutions have data on the same customers, but the data consists of different customer attributes.
- *federated transfer learning*: Both the set of customers of participating institutions and the set of customer attributes have moderate-size intersections that are large enough for smooth transfer of information from one data set to the other to enable high-quality predictions on missing customers and missing attributes.

In the remainder of this chapter, we will discuss each type of federated learning in more detail and their procedures to protect customer privacy.

Many variations and extensions of the federated learning framework have been developed over the years, e.g., [4, 33, 40, 46, 47, 58, 62, 90, 94]. Some excellent surveys on federated learning with details on recent work are [1, 7, 11, 32, 37, 41, 42, 45, 48, 57, 82, 97–99, 102]. Some recent surveys on specialized areas or applications of federated learning are [19, 20, 66, 68, 85, 93, 105, 110].

4.2 Horizontal, Cross-Device Federated Learning

The *cross-device federated learning* framework was conceived by researchers at Google, whose goal was to build an alternative to aggregating private, sensitive data of its customers in a data center for analysis [59]. Of particular concern was training an AI model with a massive amount of raw data. The researchers proposed using

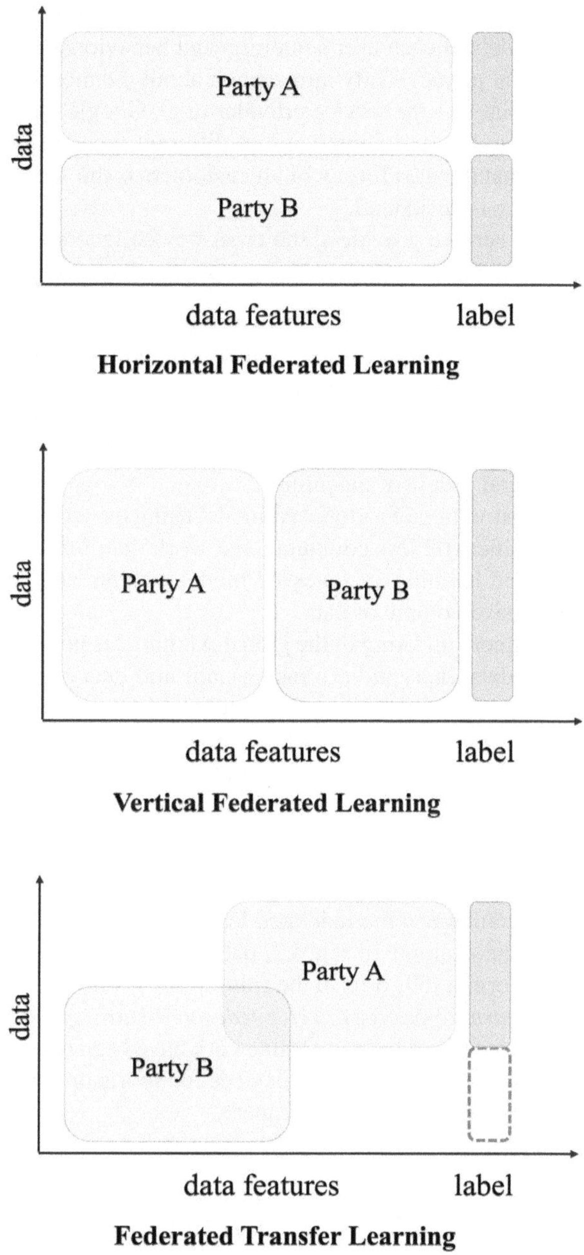

Fig. 4.1 Three types of federated learning: relationship of training data of participating institutions. In *horizontal federated learning*, participating institutions have data on different customers. The data consists of the same customer attributes. In *vertical federated learning*, participating institutions have data on the same customers, but the data consists of different customer attributes. In *federated transfer learning*, both the set of customers of participating institutions and the set of customer attributes have moderate-size intersections that are large enough for smooth transfer of information from one data set to the other to enable high-quality predictions on missing customers and missing attributes

information on personal, mobile devices (e.g., cell phones, tablets, PCs) to build local AI models of each user's interests and behaviors on the user's own device to protect personal privacy. Only information about the model (e.g., weights) is encrypted and transmitted to the service provider (e.g., Google) to be used for training a global AI model to build a global view of different types of user behaviors and preferences. Note that the data format of all customers is the same, so that cross-device federated learning is horizontal.

For service providers, the cross-device federated learning framework (shown in Fig. 4.2) has many advantages over the basic multiparty computation framework (described in Sect. 2.2):

- Multiparty computation requires encryption, transmission, and decryption of the entire massive set of raw data from each participant, while federated learning requires the same work for only the weights of the AI model. Encryption, transmission, and decryption work required for the frameworks typically differ by several orders of magnitude.
- Construction of a global AI model using the federated learning framework requires significantly less computational work than for multiparty computation since federated learning averages AI model weights and does not require training with a massive volume of data.
- Frequent updating of the global AI model is possible with federated learning since the data encryption, transmission, and decryption costs are comparatively very low, and computation of the global AI model is also considerably lower.
- The federated learning framework can use quantum-safe encryption methods (that require significantly more computing power than conventional methods) to encrypt the weights of the AI model since the volume of sensitive data that is transmitted is very low.

Figure 4.3 is an overview diagram illustrating the differences between the multiparty computation and the federated learning frameworks.

In the example in Fig. 4.2, data from all participants (cell phone users) is of the same format [59, 60], so the framework is known as *horizontal, cross-device federated learning* or *cross-device federated learning*. Many variations and applications of this federated learning framework have been developed. Other common examples are: IoT devices; wearable devices for sports or health monitoring; and smart home devices, such as vacuum cleaners.

4.3 Cross-Silo Federated Learning

Another commonly used horizontal federated learning framework is *cross-silo federated learning*, also known as *data-partitioned federated learning* [32, 40, 44]. In contrast to cross-device federated learning, which involves aggregating data summaries from millions of mobile devices by one company (such as Google, LINE, or

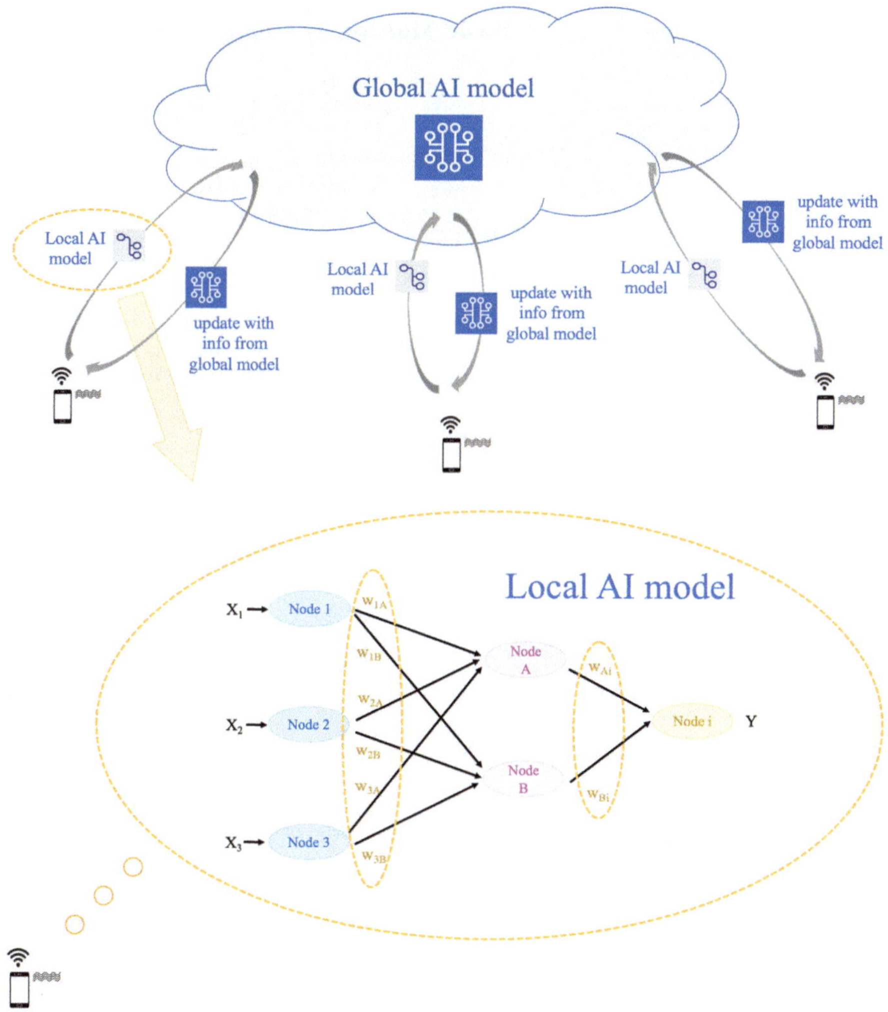

Fig. 4.2 Cross-device federated learning. Each mobile device trains an AI model locally using personal, private data on the user's mobile device. Only information about the AI model, such as the model weights $\{w_{ij}\}$, is shared with other parties. User data *never* leaves the user's device. A global AI model is trained by (weighted or straightforward) averaging local AI model weights from mobile device owners. The global AI model yields an accurate and robust AI model to predict information that is more likely to be relevant to users in relatively small demographics. Cell phone service providers can increase their revenue by increasing relevancy of information and advertisements displayed on a customer's phone

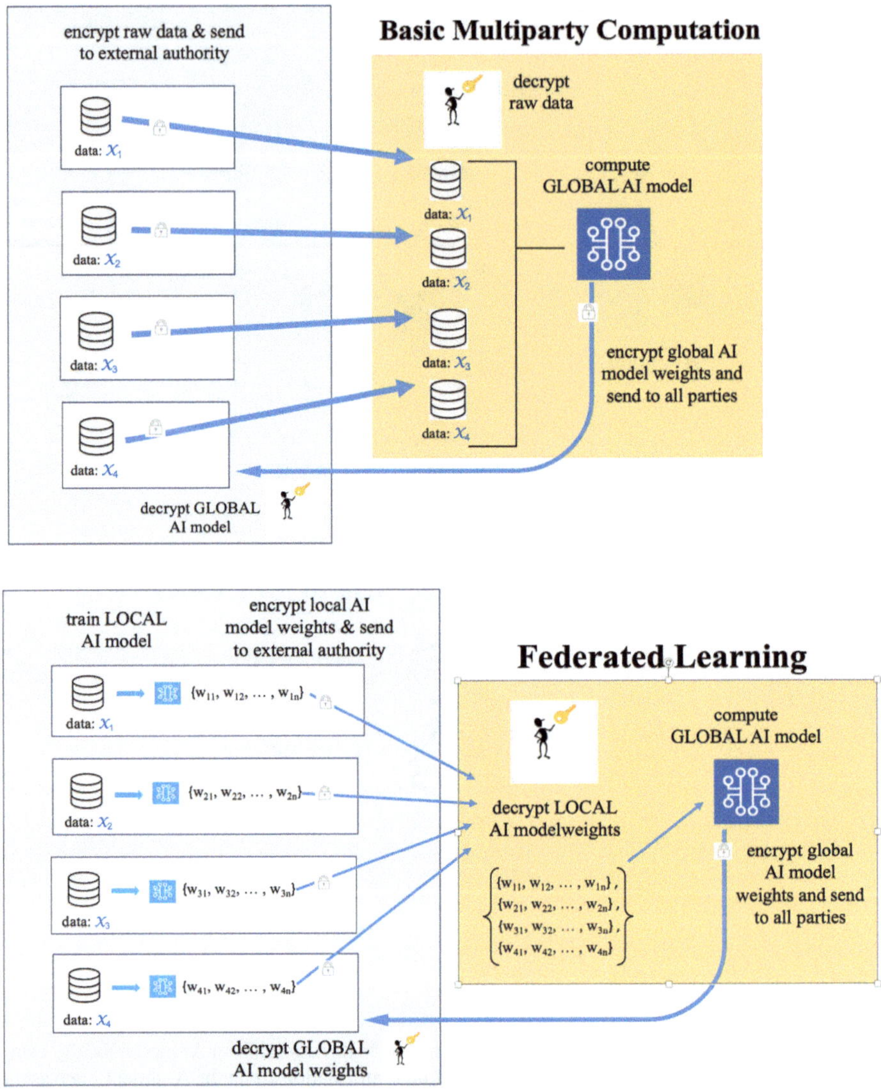

Fig. 4.3 Basic multiparty computation versus federated learning: data encryption, transmission, decryption, and computational costs. Multiparty computation requires encryption, transmission, and decryption of the entire massive set of raw data from each participant to the external authority, while federated learning requires the same work for only the weights of a local AI model. Encryption, transmission, and decryption work required for the frameworks typically differ by several orders of magnitude. Construction of a global AI model using the federated training framework requires significantly less computational work than for multiparty computation since federated learning averages AI model weights and does not require training with a massive volume of data. Frequent updating of the global AI model is possible with federated learning since data encryption, transmission, and decryption costs are comparatively very low, and computation of the global AI model is also very low. The federated learning framework can use quantum-safe encryption methods (that require significantly more computing power than conventional methods) to encrypt the weights of the AI model since the volume of sensitive data that is transmitted is very small

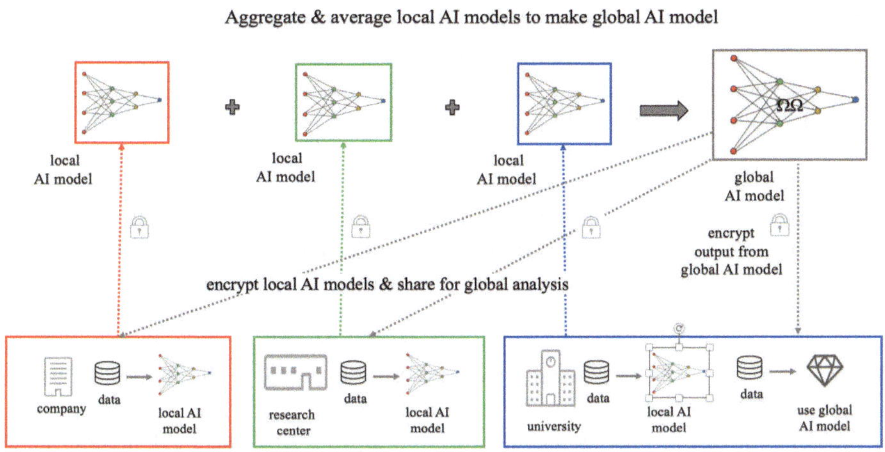

Fig. 4.4 Horizontal, cross-silo federated learning. Cross-silo federated learning typically involves collaboration between several different institutions with similar data. However, each participating institution has an insufficient volume of high-quality data for training an AI model. The underlying idea is that learning from the collective knowledge and information in all of the participants' data sets is likely to lead to a more accurate and robust global AI model than from any single local model. The first, key step is to find at least one AI model that shows promise in yielding valuable insights and computing accurate predictions. Information about the AI model can be used to convince prospective participants about the values of working together, in a federated learning framework. Each institution will train the AI model on-premise, using its own private data. The output will be a local model that is based on local data. A secure method must be established for transmitting local AI model weights $\{w_{ij}\}$ to an external, central authority for computing a global AI model. Note that raw data will never leave a participating institution's premises. A global AI model is trained by (weighted or straightforward) averaging local AI model weights from all participants. The global AI model is shared with all participants

Apple), cross-silo federated learning typically involves several different institutions with similar types of personal data. The institutions are interested in participating in federated learning since each has an insufficient volume of high-quality data for training an AI model, as shown in Fig. 4.4. The underlying idea is that learning from the collective knowledge and information in all of the institutions' data sets is likely to lead to a more accurate and robust global AI model than from any single local model. Typical scenarios involve mid-size institutions with insufficient training data or highly skewed data—such as transactions at financial institutions with a very small number of fraudulent transactions, or medical data with a very small number of patients with a rare disease.

The involvement of multiple institutions, each of whom would like to preserve privacy of their data contents, introduces additional steps to the cross-device horizontal federated learning model discussed in the previous section.

- The first, key step is to find one (or a few) AI model(s) that show promise in yielding valuable insights and accurate predictions.

- Information about the AI model(s) can be used to convince prospective participants to agree to work together, following a federated learning framework.
- Each participating institution will train the AI model(s) on-premise, using its own private data. The output will be a local model that is based on local data.
- Sometimes, the intermediate gradients from local models are shared with the trusted external authority (not other participants) to speed up training of the global AI model, particularly for data sets that require massive computations (such as images) until convergence. Recently, this practice has become less common since gradients can sometimes cause inadvertent leakage of information about raw data, so special encryption and security measures are needed to ensure protection of gradients.
- The weights $\{w_{ij}\}$ of the local AI model will be shared with other parties. Raw data never leaves a participant's premises.
- A secure method must be established for transmitting local AI models to an external, central authority, and for training thereafter.

A noteworthy point is that most real-life deployments accommodate some form of asynchronous transmission, in which participants do not coordinate with one another, i.e., one participant may send information several times before others have submitted anything. And loss of data during transmission or temporary glitches is common, so it is unrealistic to expect all participants to send summary information at the same time.

Finding a good AI model for the problem at-hand may not be easy if none of the participating institutions has even close to a sufficient volume of data for testing AI models to eliminate unpromising candidates. When all institutions have only small data sets, it may be necessary to find an external authority to aggregate encrypted, raw data to test several AI models. A major concern for participating institutions is that a cross-silo federated learning framework requires highly reliable data security to prevent leakage of private, sensitive data between participants [55]. Some possible approaches to security include

- trusted execution environment[1] [39, 81];
- differential privacy [14, 15, 90]; and
- homomorphic encryption [21, 22, 27, 74, 88, 89].

A *trusted execution environment* (TEE) consists of a hardware isolation mechanism with a secure operating system. Sometimes, a TEE may not be an option due to the need for specialized hardware with a specialized operating system. TEEs are often used in high-grade mobile devices. A famous example is the secure enclave built into iPhones.

The aim of *differential privacy* is to provide statistical or summary information about a group of individuals in a data set, without revealing any personal information

[1] Secure Technology Alliance Mobile Council White Paper: Trusted Execution Environment (TEE) 101: A Primer (June 2018): www.securetechalliance.org/wp-content/uploads/TEE-101-White-Paper-V1.1-FINAL-June-2018.pdf.

about a particular individual. Differential privacy adds some noise to data to make precise decryption difficult or impossible in a *reconstruction attack*. It is relatively fast and secure. Dinur and Nissim [12] showed that information in a database can be revealed through a small number of random queries. Their foundational work was refined and further developed by Dwork, Roth, et al. [14, 15], who established the *Fundamental Law of Information Recovery*, which shows that noise must be inserted during analysis of sensitive data to preserve and protect individual privacy. Depending on the data, the amount of noise required to prevent identification of an individual's sensitive information may be so large that the degradation in accuracy may be unacceptable. For example, in medical and financial applications, even a relatively small amount of noise is unacceptable.

Homomorphic encryption[2] is a post-quantum secure encryption method that is currently the most secure and accurate option [21, 22, 27, 74, 88, 89]. However, it is very computationally expensive and time-consuming to use, in practice. One saving grace is its support of additive and/or multiplicative properties of encrypted data. More precisely, an encryption operator E is *homomorphic* with respect to an operator $\circledcirc$ if:

$$E(x) \circledcirc E(y) = E(x \circledcirc y), \ \forall x, y \in \mathbb{M}, \tag{4.1}$$

where $\mathbb{M}$ denotes a set of plain text.

Fully homomorphic encryption algorithms support both additive and multiplicative properties on encrypted data, while *partially homomorphic encryption* algorithms are less computationally expensive to implement, but only support either additive or multiplicative operations on encrypted data (not both). Homomorphic encryption introduces a very tiny amount of error each time it is used, so mathematical operations can only be applied to the data a very large, but finite number of times. For a quick introduction to the topic for laymen, see [78].

The idea behind homomorphic encryption was originally proposed by Rivest, Adleman, and Dertouzos [74], but the first to present a plausible method to implement the idea for real-world deployment was Gentry [21]. Currently, homomorphic encryption can only be applied to very small sets of numerical or categorical data, such as weights of an AI model. It is impossible in the foreseeable future to use it to encrypt and decrypt raw text data in massive data sets typically owned by many institutions.

Zhang [107] proposed a very fast variation of homomorphic encryption specifically designed for (non-text data) in cross-silo federated learning frameworks that can reduce the computational cost while adding only a percentage or so of error. It may be useful in application scenarios that can tolerate small errors.

The federated learning models in Figs. 4.2 and 4.4 assume that the distributions of data of all participating parties are similar. In many real-world cases, the local models of individual participants are quite different. Although results from the global AI model yield an interesting look into differences with competitors, participants also want to improve their own local model with additional data. Recent works to improve

[2] Homomorphic Encryption, References: https://people.csail.mit.edu/vinodv/FHE/FHE-refs.html.

local models while protecting the privacy of the data of each of the participating institutions have shown promise [24, 30, 44, 49, 67].

The example shown in Fig. 4.4 assumes that information on local models is transmitted to an external authority for construction of a global model. In many real-world deployments, prospective participants may feel uncomfortable with dependence on an external party or it may be difficult to reach unanimous consent on a suitable external authority. As in the case with multiparty computation (see Chap. 2), a decentralized federated learning framework can be implemented with participants directly sending each other information in encrypted form, eliminating the need for an external authority [76, 106]. Two formats that are often used in deployments are

- *cyclic transfer*, in which participants send information to each other via a circular chain and
- *random transfer*, in which participants randomly select another participant to receive information about the AI model.

In both protocols, the transfers end when the termination condition is met, e.g., convergence condition for the global AI model is fulfilled.

4.4 Vertical Federated Learning

A third commonly used framework is *vertical federated learning* also known as *feature-partitioned federated learning*, in which different institutions have different data about the same customers [8, 16, 28, 38, 51, 91, 96, 97, 111]. Early vertical federated learning frameworks were designed for only two parties [28, 96], but they have since been extended to a multiparty, multi-class framework [16].

A typical application scenario involves an online e-commerce company, bank and social media company, all of whom find it beneficial to share information about their data in exchange for additional information. The e-commerce company would like to increase product recommendations that lead to customer purchases using social media "likes" data and information from demographic analysis of banking data. The social media company would like to push content that is more likely to be viewed and keep users hooked on their site, based on analysis of purchase patterns and banking data. The bank would like to increase the accuracy of its credit score for its customers based on their behaviors on social media (posts, views, comments, emojis, etc.) and purchase patterns on e-commerce sites.

In vertical federated learning, the features of interest are distributed among the participating parties, and some or all of the objects with the features appear in the raw data of participating parties. Although details in implementation and deployment may vary, the following two basic steps appear in vertical federated learning systems:

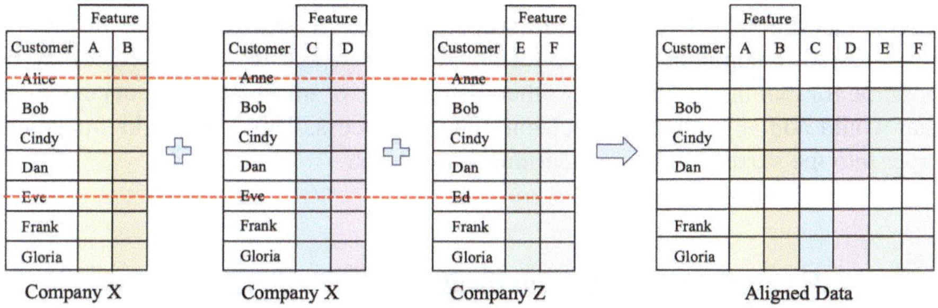

Fig. 4.5 Vertical federated learning: user alignment. Identify the set of users that appear in all of the local data sets using a secure user alignment procedure, such as *private set intersection* or *private set union*. Cell phone numbers and e-mail addresses are among the reliable attributes to spot a person who appears in the databases of different participating parties

- Step 1: *User alignment*: Identify the set of users that appear in all of the local data sets (example shown in Fig. 4.5).
- Step 2: *Model training*: Collaboratively train a global neural network using method that splits a neural network into segments and share information between segments to collaboratively train an AI model.

In the first step, user alignment can be carried out using *private set intersection*, a secure multiparty computation framework that enables parties to compute the intersections of their sets (e.g., customers, employees) without revealing either party's individual set [61, 75]. A more recent method, *private set union*, computes the union of data sets and outputs only the final result [18, 23, 35, 36]. Private set union appears to be more secure regarding the membership leakage problem. Cell phone numbers and e-mail addresses are among the reliable attributes to spot a person who appears in the databases of different participating parties.

In the second step, two commonly used approaches to collaboratively train a global model are SplitNN and FedBCD. Both methods require sharing of intermediate results during the training process. SplitNN involves splitting a neural network model into different segments. Each party can individually train a segment, and intermediate results are shared after each step of training a segment in order to collaboratively train a global model [13, 26]. In FedBCD (federated block coordinate descent), several rounds of training take place individually before sharing results to significantly reduce the communication overhead [50, 52].

Vertical federated learning has two main architectures: one with and without an external trusted authority. The initial framework proposed in [28] combines information (e.g., attributes of a customer) from two parties while relying on an external coordinator. The external coordinator is responsible for generating keys to ensure privacy and compute gradients and the loss during training [98]. This framework was extended to combine information from two parties without the need for an external coordinator in subsequent works [29, 83, 101]. The framework was further extended to enable sharing of attributes by parties of three or more [9, 109].

In some real-world scenarios of vertical federated learning, only one institution is interested in supplementing the features of its data set. In such cases, it may be possible for that institution to pay other institutions for information about customers that would add value to their decision-making process. This would add a payment layer into the vertical federated learning framework.

4.5 Federated Transfer Learning

The *federated transfer learning* framework incorporates ideas from both federated learning and transfer learning. This section begins with a quick introduction to transfer learning before describing a simple example of federated transfer learning involving two parties. We will not be touching upon multiparty federated transfer learning, a much more difficult problem that is fast becoming a hot research topic.

4.5.1 Transfer Learning

The field of *transfer learning* came into fruition as increasingly large data sets were required for implementing data mining and machine learning algorithms[3] [5, 31, 53, 63, 64, 71, 73, 77, 100, 112]. In the business world, environmental conditions are perpetually undergoing change, so the feature space evolves over time. For example, the distribution of features and the features themselves may change, or new features may emerge and others may disappear. For time-series data, transferring knowledge from one time window to the next, then fine-tuning machine learning and AI models to take into account changes would likely converge faster (minimize computation and training time) and yield more valuable business insights than treating each time window in isolation, as an independent unit.

The application of transfer learning to scenarios involving temporal evolution of an institution's data set has evolved into an independent field of research, known as *continual lifelong learning*, and more effective approaches to deal with the problem are being developed [65].

Transfer learning is useful when very little data from a source is available for building an AI model, but massive sets of data from the same domain are available. Transfer learning is used to build an AI model with the massive data sets, then the AI model is reused (or repurposed) as a starting point for a similar task with the smaller data set that is of interest. In this scenario, transfer learning involves the following steps:

[3] NIPS'95 Post-Conference Workshop: Learning to Learn: Knowledge Consolidation and Transfer in Inductive Systems, website: https://socrates.acadiau.ca/courses/slashcomp/dsilver/NIPS95_LTL/transfer.workshop.1995.html.

- Train an AI model using the massive data set.
- Freeze all of the weights of the AI model, except the final layer.
- Replace the final layer of the AI model with one that is appropriate for the purposes of analysis with the small data set.
- Fine-tune the AI model for the smaller data set by training just the final layer of the AI model using the smaller data set as the training data.

Transfer learning is known to work well for image data sets because the first several layers of deep neural networks are known to identify general, high-level features, such as edges and textures. Only the final or final few layers are highly tuned for a specific task.[4,5,6] Since the number of final layers that are effective at task-specific feature identification are unknown, if the smaller data set consists of a moderate amount of data, e.g., tens thousands of images, then removing the final two or three layers and training the new AI model may yield better results than replacing only the final layer. In general, implementation studies are needed to determine the number of layers to replace to obtain the best AI model.

Transfer learning is easy to implement in domains for which large, open data sets are available. For example, there are many large repositories of high-quality, open image data sets, such as *ImageNet*.[7] Some researchers and organizations[8,9,10] have trained highly sophisticated AI models on these open data sets, and have made the AI models available. To facilitate transfer learning, some open AI model freewares have even added a simple option of freezing the weights of all nodes except those in the final output layer. Users can replace the final output layer with one that meets their own needs, as shown through a mock example in Fig. 4.6. Training large AI models with massive data sets can take many days and have high costs associated with labor and computer time, so it may be useful to try transfer learning using a publicly available AI model that has already been trained on a famous benchmark data set.[11]

An important caveat regarding transfer learning is that it does not always work; the new AI model may look fine, but may be useless junk. Testing the model is of utmost importance. Reasons for failure are often unclear and may be difficult to pinpoint. In some instances, the properties of the large data set used to train an AI model may not be sufficiently similar to the properties of the smaller data set that will use the

---

[4] Ryan Ahmed, What is Transfer Learning?, video lecture: www.youtube.com/watch?v=3gyeDlZqWko/

[5] Derrik Mwiti, Transfer Learning Guide: A Practical Tutorial With Examples for Images and Text in Keras, MLOps Blog, April 22, 2024: http://neptune.ai/blog/transfer-learning-guide-examples-for-images-and-text-in-keras.

[6] Sebastian Ruder, Transfer Learning, Machine Learning's Next Frontier, Mar 21, 2017: www.ruder.io/transfer-learning.

[7] ImageNet website: www.image-net.org.

[8] Google Teachable Machine FAQ website: http://teachablemachine.withgoogle.com/faq.

[9] Google MediaPipe Model Maker website: http://ai.google.dev/edge/mediapipe/solutions/model_maker.

[10] NVIDIA TAO Toolkit website: http://developer.nvidia.com/tao-toolkit.

[11] Andrew Ng, Transfer Learning, video lecture: www.youtube.com/watch?v=FQM13HkEfBk.

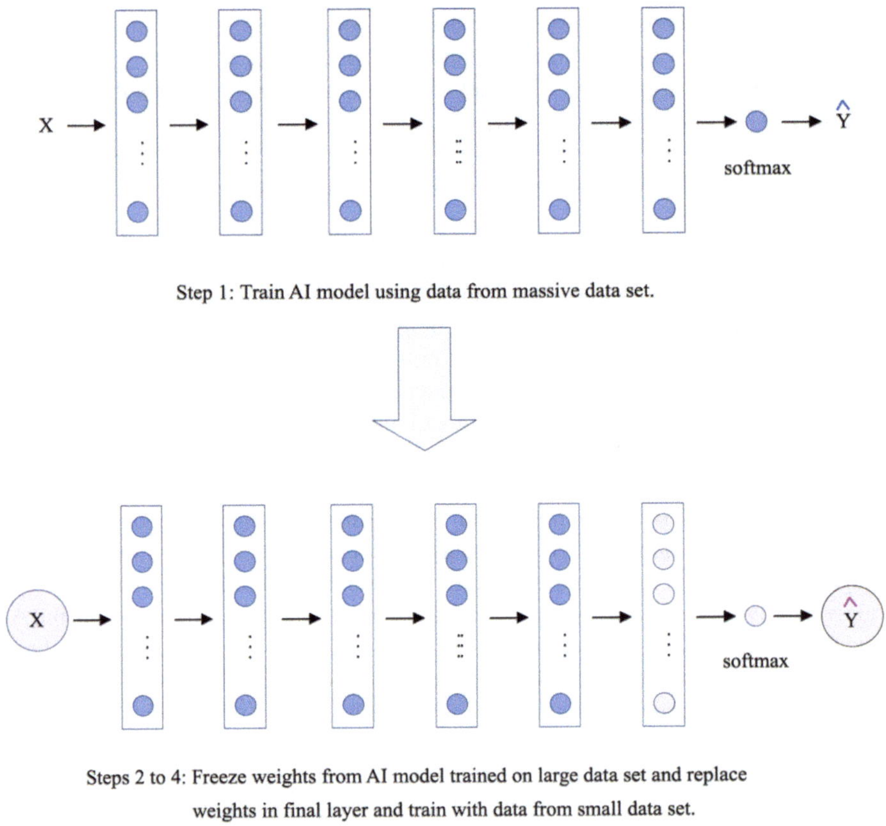

Fig. 4.6 Transfer learning example. A multi-layer, AI model trained on a massive, open image data set can be fine-tuned for use in a specialized purpose, such as medical image analysis for diagnosis of diseases. **Step 1:** Train a multi-layer neural network using a massive image data set, such as *ImageNet*, or use a publicly available AI model that has been trained on a famous open image data set. **Step 2:** Freeze all of the weights of the AI model, except the final layer. **Step 3:** Replace the last layer of the AI model with one that is appropriate for the purposes of analysis with the small data set. **Step 4:** Fine-tune the AI model for the smaller data set by training just the final layer of the AI model

transferred AI model. For example, a special property of the smaller data set that is vital for correct and reliable classification may not exist in the large data set. Another possible cause for failure is that the AI model may not have the correct architecture to capture a special property of the smaller data set that is vital for classification.

In any prospective transfer learning scenario, results should be checked through implementation and testing with the data sets and AI model. In highly sensitive applications, such as medical diagnosis, transfer learning should only be considered as an additional tool that may provide a helpful second opinion, not a replacement for a highly trained medical professional.

Due to its widespread applicability, transfer learning has appeared in many different scenarios under different guises, including: *learning to learn, lifelong learning, knowledge transfer, inductive transfer, multitask learning, knowledge consolidation, context-sensitive learning, knowledge-based inductive bias, metalearning,* and *incremental/cumulative learning* [53, 63, 64, 86, 87, 100, 112].

4.5.2 Federated Transfer Learning: A Two-Party Example

As mentioned earlier, federated transfer learning incorporates ideas from both federated learning and transfer learning. Consider the following example of a transfer learning framework involving only two parties. Suppose the two parties, A and B, have data sets that partially overlap, as shown in the bottom of Fig. 4.1. All of Party A's data is labeled, while only part of Party B's data is labeled. More specifically, only the data in B that partially overlaps with A is labeled. The goal is to predict the remaining labels for Party B's data.

The algorithm for solving this two-party problem is as follows:

- Train an AI model using Party A's data set.
  - Randomly split Party A's data set into a training set and test set; the ratio is typically 80–20%.
  - Use the training data to train the AI model.
  - Verify that the AI model architecture is appropriate using the test set.
  - If the errors are unacceptably large, try different AI model architectures until one with high-quality predictions is found.
- Use transfer learning to train the AI model from A to create an AI model for Party B's data.
  - Remove the final output layer in the AI model from Party A and replace it with an appropriate layer for Party B.
  - Train the modified AI model.
  - Verify that the AI model architecture is appropriate using the labeled data from Party B.

Since only the AI model architecture and AI model weights are transferred from Party A to Party B (i.e., no raw data is transferred or revealed by either party), this algorithm is considered to be a type of federated learning, hence the name federated transfer learning.

As mentioned earlier, transfer learning is not always successful. Federated transfer learning may be unsuccessful for similar reasons. Depending on the similarities and differences of properties of the data from Parties A and B, transfer learning may or may not be successful. For example, an architecture that is appropriate for finding distinguishing characteristics in data elements from Party A may not be appropriate for distinguishing data elements from Party B. The size of the overlap between the data sets may also affect the outcome. In short, federated transfer learning is one of many tools that can be tried out to determine whether it can yield a reliable AI model for prediction [103].

Many variations and examples of the framework are examined in reviews of federated transfer learning [5, 25, 31, 34, 43, 53, 63, 64, 71, 73, 77, 84, 86, 95, 100, 104, 112]. In particular, Chap. 4 in [84] examines federated transfer learning, in which one party has data that is partly or entirely multi-modal (i.e., data with at least two modalities) and other parties hold data that is partly or entirely uni-modal (i.e., data with only one modality). The idea is to have the party with multi-modal data help the parties with uni-modal data while protecting the privacy of the data of all parties.

References

1. Abdulrahman, S., Tout, H., Ould-SLimane, H., Mourad, A., Talhi, C., Guizani, M.: A survey on federated learning: the journey from centralized to distributed on-site learning and beyond. IEEE Inter. Things J. **8**(7), 5476–5497 (2021). arXiv:1902.01046. Accessed 10 June 2024
2. Aledhari, M., Razzak, R., Parizi, R.M., Saeed, F.: Federated learning: a survey on enabling technologies, protocols, and applications. IEEE Access **8** (2020) . https://par.nsf.govservlets/purl/10221843. Accessed 8 March 2024
3. Bonawitz, K.,Eichner, H., Grieskamp, W., Huba, D., Ingerman, A., Ivanov, V., Kiddon, C., Konecny, J., Mazzocchi, S., McMahan, H.B., et al.: Towards federated learning at scale: System design (2019). arXiv:1902.01046. Accessed 10 April 2024
4. Bonawitz, K.A., Ivanov, V., Kreuter, B., Marcedone, A., McMahan, H.B., Patel, S., Ramage, D., Segal, A., Seth, K.: Practical secure aggregation for federated learning on user-held data. In: Proceedings of Neural Information Processing Systems Workshop on Private Multi-Party Machine Learning. arXiv:1611.04482. Accessed 24 June 2024
5. Chato, L., Regentova, E.: Survey of transfer learning approaches in the machine learning of digital health sensing data. J. Med. **13**(12), 1703 (2023). www.mdpi.com/2075-4426/13/12/1703. Accessed 11 March 2024
6. Che, L., Wang, J., Zhou, Y., Ma, F.: Multimodal federated learning: a survey. Sensors **23**(15), 6986 (2023). https://doi.org/10.3390/s23156986. Accessed 19 March 2024
7. Chen, H., Wang, H., Long, Q., Jin, D., Li, Y.: Advancements in federated learning: Models, methods, and privacy. ACM Comput. Surv. (2024). https://dl.acm.org/doi/10.1145/3664650. Accessed 18 June 2024
8. Chen, P., Du, X., Lu, Z., Wu, J., Hun, P., Hun, C.K.: EVFL: an explainable vertical federated learning for data-oriented artificial intelligence systems. J. Syst. Arch. **126**, 102474 (2022). www.sciencedirect.com/science/article/abs/pii/2024.S1383762122000583. Accessed 10 April 2024
9. Cheng, K., Fan, T., Jin, Y., Liu, Y., Chen, T., Papadopoulos, D., Yang, Q.: Secureboost: a lossless federated learning framework. IEEE Intell. Syst. **36**(6), 87–98 (2021)
10. Chilimbi, T., Suzue, Y., Apacible, J., Kalyanaraman, K.: Project Adam: building an efficient and scalable deep learning training system. In: USENIX Symposium on Operating Systems Design and Implementation (2014). www.usenix.org/system/files/conference/osdi14/osdi14-paper-chilimbi.pdf. Accessed 29 May 2024
11. Ding, J., Tramel, E., Kumar Sahu, A.K., Wu, S., Avestimehr, S., Zhang, T.: Federated learning challenges and opportunities: an outlook, pp. 8752–8756. In: Proceedings of IEEE International Conference on Acoustics, Speech and Signal Processing (ICASSP) (2022)

12. Dinur, I., Nissim, K.: Revealing information while preserving privacy. In: Proceeding of ACM SIGMOD-SIGACT-SIGART Symposium on Principles of Database Systems (PODS), pp. 202–210 (2003). https://al.acm.org/doi/10.1145/773153.773173. Accessed 18 April 2024
13. Duan, Q., Hy, S., Deng, R., Lu, Z.: Combined federated and split learning in edge computing for ubiquitous intelligence in internet of things: state-of-the-art and future directions. Sensors (Basel) **22**(16), 5983 (2022). www.ncbi.nlm.nih.gov/pmc/articles/PMC9414384. Accessed 14 June 2024
14. Dwork, C., McSherry, F., Nissim, K., Smith, A.: Calibrating noise to sensitivity in private data analysis. In: Proceedings of Theory of Cryptography Conference, pp. 265–284. Springer, Berlin, Germany (2006)
15. Dwork, C., Roth, A.: The algorithmic foundations of differential privacy. Found. Trends Theor. Comput. Sci. **9**(3–4), 211–407 (2014). https://al.acm.org/doi/10.1561/0400000042. Accessed 18 April 2024
16. Feng, S., Yu, H.: Multi-participant multi-class vertical federated learning (2020). arXiv:2001.11154. Accessed 18 June 2024
17. Gabrielli, E., Pica, G., Tolomei, G.: A survey on decentralized federated learning (2023). arXiv:2308.04604. Accessed 8 March 2024
18. Gao, J., Nguyen, S., Trieu, N.: Toward a practical multi-party private set union. Cryptology ePrint Archive (2024). https://eprint.iacr.org/2023/1930.pdf. Accessed 14 June 2024
19. Gecer, M., Garbinato, B.: Federated Learning for Mobility Applications. ACM Comput. Surv. **56**(5) 133, 1–28 (2024). https://dl.acm.org/doi/10.1145/3637868. Accessed 18 June 2024
20. Geng, S., Yu, H.: Multi-participant multi-class vertical federated learning (2020). arXiv:2001.11154. Accessed 18 June 2024
21. Gentry, C.: A fully homomorphic encryption scheme. Ph.D. Thesis, Department of Computer Science, Stanford University. https://crypto.stanford.edu/craig/craig-thesis.pdf. Accessed 18 April 2024
22. Gentry, C.: Computing arbitrary functions of encrypted data. Commun. ACM **53**(3), 97–105 (2010). https://al.acm.org/doi/10.1145/1666420.1666444. Accessed 18 April 2024
23. Gong, X., Hua, Q.-S., Jin, H.: Nearly optimal protocols for computing multi-party private set union. In: Proceedings of IEEE/ACM International Symposium on Quality of Service (IWQoS), pp. 1–10 (2022)
24. Gorbett, M., Shirazi, H., Ray, I.: Cross-silo federated learning across divergent domains with iterative parameter alignment. In: Proceedings of IEEE International Conference on Big Data (2023). arXiv:2311.04818. Accessed 15 April 2024
25. Guo, W., Zhuang, F., Zhang, X., Tong, Y., Dong, J.: A comprehensive survey of federated transfer learning: challenges, methods and applications (2024). arXiv:2403.01387. Accessed 15 April 2024
26. Gupta, O., Raskar, R.: Distributed learning of deep neural network over multiple agents. J. Netw. Comput. Appl. **116**, 1–8 (2018)
27. Halevi, S.: Homomorphic Encryption. Tutorial on the Foundations of Cryptography. In: Lindell, Y. (ed.), Tutorials on the Foundations of Cryptography. Springer, Berlin, Germany (2017). https://shaih.github.io/pubs/he-chapter.pdf. Accessed 18 April 2024
28. Hardy, S., Henecka, W., Ivey-Law, H., Nock, R., Patrini, G., Smith, G., Thorne, B.: Private federated learning on vertically partitioned data via entity resolution and additively homomorphic encryption (2017). arXiv:1711.10677. Accessed 10 April 2024
29. He, D., Du, R., Zhu, S., Zhang, M., Liang, K., Chan, S.: Secure logistic regression for vertical federated learning. IEEE Internet Comput. **26**(2), 61–68 (2021)
30. Heikkilä, M.A., Koskela, A., Shimizu, K., Kaski, S., Honkela, A.: Differentially private cross-silo federated learning (2020). arXiv:2007.05553. Accessed 15 April 2024
31. Hosna, A., Merry, E., Gyalmo, J., Alom, Z., Aung, Z., Abdul Azim, M.: Transfer learning: a friendly introduction. J. Big Data **9**, 102 (2022). https://journalofbigdata.springeropen.com/articles/10.1186/s40537-022-00652-w. Accessed 16 March 2024
32. Huang, C., Huang, J., Liu, X.: Cross-silo federated learning: challenges and opportunities (2022). arXiv:2206.12949. Accessed 19 March 2024

33. Huang, W., Li, T., Wang, D., Du, S., Zhang, J., Huang, T.: Fairness and accuracy in horizontal federated learning. Inf. Sci. **589**, 170–185 (2022). https://zhangjunbo.org/pdf/2022_InsSci_FedFa.pdf. Accessed 8 April 2024
34. Iman, H., Arabnia, H.R., Rasheed, K.: A review of deep transfer learning and recent advancements. Technologies **11**(2), 40 (2023). www.mdpi.com/2227-7080/11/2/40. Accessed 26 June 2024
35. Jia, Y., Sun, S.-F., Zhou, H.-S., Du, J., Gu, D.: Shuffle-based private set union. Faster and more secure. In: Proceedings of USENIX Security Symposium (2022). www.usenix.org/system/files/sec22-jia.pdf. Accessed 10 June 2024
36. Jia, Y., Sun, S.-F., Zhou, H.-S., Gu, D.: The ideal functionalities for private set union. Cryptology ePrint Archive (2022). https://ia.cr/2022/750. Accessed 10 June 2024
37. Kairouz, P., McMahan, B., Avent, B., Bellet, A., Bennie, M. et al.: Advances and open problems in federated learning. Found. Trends Mach. Lear. **14**(1–2), 1–210 (2021). arXiv:1912.04977. Accessed 10 April 2024
38. Khan, A., Thij, M.T., Wilbik, A.: Vertical Federated learning: a structured literature review (2022). arXiv:2212.00622. Accessed 8 April 2024
39. Lee, D.: Building Trusted Execution Environments. Ph.D. Dissertation, Department of Computer Science, University of California at Berkeley (2022). www2.eecs.berkeley.edu/Pubs/TechRpts/2022/EECS-2022-96.pdf. Accessed 10 May 2024
40. Li, L., Fan, Y., Tse, M., Lin, K.-Y.: A review of applications in federated learning. Comput. Ind. Eng. **149**, 106854 (2020). https://orca.cardiff.ac.uk/id/eprint/134968. Accessed 20 May 2024
41. Li, Q., Wen, Z., Wu, Z., Hu, S., Wang, N., Li, Y., Liu, X., He, B.: A survey on federated learning systems: vision, hype and reality for data privacy and protection. IEEE Trans. Knowl. Data Eng. **35**, 3347–3366 (2023). arXiv:1907.09693. Accessed 19 March 2024
42. Li, T., Kumar Sahu, A., Talwalkar, A., Smith, V.: Federated learning: challenges, methods, and future directions. IEEE Signal Proc. Mag. **37**(3), 50–60 (2020). arXiv:1908.07873. Accessed 4 June 2024
43. Li, W., Huang, R., Li, J., Liao, Y., Chen, Z., He, G., Yan, R., Gryllias, K.: A perspective survey on deep transfer learning for fault diagnosis in industrial scenarios: theories, applications and challenges. Mech. Syst. Signal Proc. **167**(A), 108487 (2022)
44. Li, Z., Chaturvedi, P., He, S., Chen, H., Singh, G., Kindratenko, V., Huerta, E. A., Kim, K., Madduri, R.: FedCompass: efficient cross-silo federated learning on heterogeneous client devices using a computing power-aware scheduler. In: Proceeding of International Conference on Learning Representations (2024). https://openreview.net/forum?id=msXxrttLOi. Accessed 15 April 2024
45. Lim, W.Y.B., Luong, N.C., Hoang, D.T., Jiao, Y., Liang, Y.C., Yang, Q., Niyato, D., Miao, C.: Federated learning in mobile edge networks: a comprehensive survey. IEEE Commun. Surv. Tutor. **3**, 2031–2063 (2020)
46. Liu, D.,Bai, L., Yu, T., Zhang, A.: Towards method of horizontal federated learning: a survey. In: Proceedings of IEEE International Conference on Big Data and Information Analytics (BigDIA), Guiyang, China, pp. 259–266 (2022)
47. Liu, F., Zheng, Z., Shi, Y., Tong, Y., Zhang, Y.: A survey on federated learning: a perspective from multi-party computation. Front. Comput. Sci. **18**, 181336 (2024). https://link.springer.com/article/10.1007/11704-023-3282-7. Accessed 8 March 2024
48. Liu, J., Huang, J., Zhou, Y., Li, X., Ji, S., Xiong, H., Dou, D.: From distributed machine learning to federated learning: a survey. Knowl. Inf. Syst. **64**(4), 885–917 (2022)
49. Liu, K., Hu, S., Wu, S., Smith, V.: On privacy and personalization in cross-silo federated learning. In: Proceedings of Advances in Neural Information Processing Systems (2022). https://openreview.net/forum?id=Oq2bdIQQOIZ. Accessed 15 April 2024
50. Liu, Y., Kang, Y., Zhang, X., Li, L., Cheng, Y., Chen, T., Hong, M., Yang, Q.: A communication-efficient collaborative learning framework for distributed features. In: Proceedings of International Workshop on Federated Learning for Data Privacy and Confidentiality, in Conjunction with NeurIPS (2019). arXiv:1912.11187. Accessed 14 June 2024

51. Liu, Y., Kang, Y., Zou, T., Pu, Y., He, Y., Ye, X., Ouyang, Y., Zhang, Y.-Q., Yang, Q.: Vertical federated learning: concepts, advances and challenges. IEEE Trans. Knowl. Data Eng. (2024). arXiv:2211.12814. Accessed 19 March 2024
52. Liu, Y., Zhang, Y., Li, L., Chen, T.: FedBCD: a communication-efficient collaborative learning framework for distributed features. IEEE Trans. Signal Proc. **70**, 4277–4290 (2022)
53. Lu, J., Behbood, V., Hao, P., Zuo, H., Xue, S., Zhang, G.: Transfer learning using computational intelligence: a survey. Knowl.-Based Syst. **80**, 14–23 (2015)
54. Ludwig, H., Baracaldo, N. (eds.): Federated Learning: A Comprehensive Overview of Methods and Applications. Springer, Berlin, Germany (2022)
55. Ma, C., Li, J., Ding, M., Yang, H.H., Shu, F., Member, Quek, T.Q.S., Poor, H.V.: On safeguarding privacy and security in the framework of federated learning. IEEE Netw. **99**, 1–7 (2020). arXiv:1909.06512. Accessed 20 April 2024
56. Mahlool, D.H., Abed, M.H.: A comprehensive survey on federated learning: concept and applications. Lecture Notes on Data Engineering and Communications Technologies (2022). arXiv:2201.09384. Accessed 8 March 2024
57. Majeed, A., Hwang, S.O.: A multifaceted survey on federated learning: fundamentals, paradigm shifts, practical issues, recent developments, partnerships, trade-offs, trustworthiness, and ways forward. IEEE Access **12**, 84643–84679 (2024)
58. Mandal, P.K., Leo, C., Hurley, C.: Horizontal federated computer vision (2023). arXiv:2401.00390. Accessed 8 March 2024
59. McMahan, B., Moore, E., Ramage, D., Hampson, S., Arcas, B.A.Y.: Communication efficient learning of deep networks from decentralized data. In: Proceeding of International Conference on Artificial Intelligence and Statistics (2017). arXiv:1602.05629. Accessed 19 March 2024
60. McMahan, B., Ramage, D.: Federated learning: collaborative machine learning without centralized training data. Google AI Blog (2017). https://ai.googleblog.com/2017/04/federated-learning-collaborative.html. Accessed 19 March 2024
61. Morales, D., Agudo, I., Lopez, J.: Private set intersection: a systematic literature review. Comput. Sci. Rev. **49**, 100567 (2023). www.sciencedirect.com/science/article/pii/S1574013723000345. Accessed 12 June 2024
62. Nguyen, D.C., Ding, M., Pathirana, P.N., Seneviratne, A., Li, J., Poor, H.V.: Federated learning for internet of things: a comprehensive survey. IEEE Commun. Surv. Tutor. **23**(3), 1622–1658 (2021). https://ieeexplore.ieee.org/stamp/stamp.jsp?arnumber=9415623. Accessed 20 April 2024
63. Niu, S., Liu, Y., Wang, J., Song, H.: A decade survey of transfer learning (2010–2020). IEEE Trans. Artif. Intell. **1**(2), 151–166 (2020)
64. Pan, S.J., Yang, Q.: A survey on transfer learning. IEEE Trans. Knowl. Data Eng. **22**(10), 1345–1359 (2010). https://ieeexplore.ieee.org/document/5288526. Accessed 16 March 2024
65. Parisi, G.I., Kemker, R., Part, J.L., Kanan, C., Wermter, S.: Continual lifelong learning with neural networks: a review (2018). arXiv:1802.07569. Accessed 8 March 2024
66. Pfeiffer, K., Rapp, M., Khalili, R., Henkel, J.: Federated learning for computationally constrained heterogeneous devices: a survey. ACM Comput. Surv. **55**(14s)334, 1–27 (2023). https://dl.acm.org/doi/10.1145/3596907. Accessed 18 June 2024
67. Qin, Z., Deng, S., Zhao, M., Yan, X.: FedAPEN (2023) Personalized cross-silo federated learning with adaptability to statistical heterogeneity. https://dl.acm.org/doi/abs/10.1145/3580305.3599344. Accessed 18 June 2024
68. Qu, Y., Uddin, M.-P., Gan, C., Xiang, Y., Gao, L., Yearwood, J. (2022) Blockchain-enabled federated learning: A survey. ACM Comput. Surv. **55**(4)70, 1–35. https://dl.acm.org/doi/10.1145/3524104. Accessed 18 June 2024
69. Rafi, T.H., Noor, F. A., Hussain, T., Chae, D.-K., Yang, Z.: (2023) A generalized look at federated learning: survey and perspectives. arXiv:2303.14787. Accessed 8 March 2024
70. Rana, N., Marwaha, H.: Role of federated learning in healthcare systems: a survey. Math. Found. Comput. (2023). www.aimsciences.org/article/doi/10.3934/mfc.2023023. Accessed 8 March 2024

71. Razavi-Far, R., Wang, B., Taylor, M.E., Yang, Q. (eds.): Federated and Transfer Learning. Adaptation, Learning, and Optimization Series, vol. 27. Springer, Berlin, Germany (2023)
72. Reddy, K.D., Gadekallu, T.R.: A comprehensive survey on federated learning techniques for healthcare informatics. Comput. Intell. Neurosci. (2023). www.ncbi.nlm.nih.gov/pmc/articles/PMC9995203. Accessed 19 March 2024
73. Ribani, R., Marengoni, M.: A survey of transfer learning for convolutional neural networks. In: Proceeding of SIBGRAPI Conference on Graphics, Patterns and Images Tutorials (SIBGRAPI-T) (2019). https://ieeexplore.ieee.org/document/8920338. Accessed 16 March 2024
74. Rivest, R.L., Adleman, L., Dertouzos, M.L.: On data banks and privacy homomorphisms. Found. Secure Comput. **4**(11), 169–180 (1978). https://evervault.com/papers/rivest-adleman-dertouzos. Accessed 18 April 2024
75. Rosulek, M.: A brief overview of private set intersection, US NIST presentation (2021). https://csrc.nist.gov/presentations/2021/stppa2-psi. Accessed 18 April 2024
76. Roy, A.G., Siddiqui, S., Pölsterl, S., Navab, N., Wachinger, C.: BrainTorrent: a peer-to-peer environment for decentralized federated learning (2019). arXiv:1905.06731. Accessed 10 June 2024
77. Saha, S., Ahmad, T.: Federated transfer learning: concept and applications. Intelligenza Artificiale **15**(1), 35–44 (2021). arXiv:2010.15561. Accessed 18 June 2024
78. Sellers, A.: Everything you wanted to know about homomorphic encryption (but were afraid to ask). Forbes (2021). www.forbes.com/sites/forbestechcouncil/2021/04/14/everything-you-wanted-to-know-about-homomorphic-encryption-but-were-afraid-to-ask. Accessed 18 April 2024
79. Shokri, R., Shmatikov, V.: Privacy-preserving deep learning. In: Proceeding of ACM SIGSAC Conference on Computer and Communications Security, pp. 1310–1321 (2015). http://cs.cornell.edu/~shmat/shmat_ccs15.pdf. Accessed 29 May 2024
80. Shokri, R., Stronati, M., Song, C., Shmatikov, V.: Membership inference attacks against machine learning models. In: Proceedings of IEEE Symposium on Security and Privacy, pp. 3–18. arXiv:1610.05820. Accessed 4 June 2024
81. Sommerhalder, M.: Trusted execution environment (Chap. 18). In: Mulder, V., et al. (Ed.) Trends in Data Protection and Encryption Technologies, pp. 95–101. Springer, Berlin, Germany (2023). https://link.springer.com/chapter/10.1007/978-3-031-33386-6_18. Accessed 11 June 2024
82. Subasi, O., Bel, O., Manzano, J., Barker, K.: The landscape of modern machine learning: a review of machine, distributed and federated learning (2023). arXiv:2312.03120v1. Accessed 11 June 2024
83. Sun, H., Wang, Z., Huang, Y., Ye, J.: Privacy-preserving vertical federated logistic regression without trusted third-party coordinator. In: Proceeding of International Conference on Machine Learning and Soft Computing, pp. 132–138 (2022). https://dl.acm.org/doi/abs/10.1145/3523150.3523171. Accessed 9 April 2024
84. Sun, Y.: Federated Transfer Learning with Multimodal Data. M.A. Thesis, Technische Universität Darmstadt, Germany (2022). arXiv:2209.03137. Accessed 11 Aug 2024
85. Supriya, Y., Gadekallu, T.R.: A survey on soft computing techniques for federated learning: applications, challenges and future directions. J. Data Inf. Qual. **15**(2), 1–28 (2023). https://dl.acm.org/doi/10.1145/3575810. Accessed 18 June 2024
86. Tan, C., Sun, F., Kong, T., Zhang, W., Yang, C., Liu, C.: A survey on deep transfer learning (2018). arXiv:1808.01974. Accessed 26 June 2024
87. Taylor, M.E., Stone, P.: Transfer learning for reinforcement learning domains: a survey. J. Mach. Learn. Res. **10**(56), 1633–1685 (2009)
88. Vaikuntanathan, V.: Computing blindfolded. New Developments in Fully Homomorphic Encryption. In: Tutorial, IEEE Symposium on Foundations of Computer Science (FOCS) (2011). https://people.csail.mit.edu/vinodv/FHE/FHE-focs-survey.pdf. Accessed 18 April 2024

89. van Dijk, M., Gentry, C., Halevi, S., Vaikuntanathan, V.: Fully homomorphic encryption over the integers. In: Advances in Cryptology – EUROCRYPT. Lecture Notes in Computer Science, vol. 6110. Springer, Berlin, Germany (2010). https://doi.org/10.1007/978-3-642-13190-5_2. Accessed 18 June 2024
90. Wei, K., Li, J., Ding, M., Ma, C., Yang, H.H., Farokhi, F., Jin, S., Quek, T.Q.S., Poor, H.V.: Federated learning with differential privacy: algorithms and performance analysis. IEEE Trans. Inf. Forens. Secur. **15**, 3454–3469 (2022). arXiv:1911.00222. Accessed 20 May 2024
91. Wei, K., Li, J., Ma, C. Ding, M., Wei, S., Wu, F., Chen, G., Ranbaduge, T.: Vertical federated learning: challenges, methodologies and experiments (2022). arXiv:2202.04309. Accessed 20 May 2024
92. Wen, J., Zhang, Z., Lan, Y., Cui, Z., Cai, J., Zhang, W.: A survey on federated learning: challenges and applications. Int. J. Mach. Learn. Cybern. **14**, 513–535 (2023). Published online 11 Nov 2022. https://link.springer.com/article/10.1007/s13042-022-01647-y. Accessed 20 May 2024
93. Wu, J., Dong, F., Leung, H., Zhu, Z., Zhou, J., Drew, S.: Topology-aware federated learning in edge computing: a comprehensive survey. ACM Comput. Surv. **56**(10), 1–41 (2024). https://dl.acm.org/toc/csur/2024/56/10. Accessed 18 June 2024
94. Wu, W.: Towards efficient horizontal federated learning. Ph.D. Thesis, Department of Computer Science, University of Warwick (2021). https://wrap.warwick.ac.uk/id/eprint/165407/. Accessed 20 May 2024
95. Yan, P., Abdulkadir, A., Luley, P.-P., Rosenthal, M., Schatte, G.A., Grewe, B.F., Stadelmann, T.: A comprehensive survey of deep transfer learning for anomaly detection in industrial time series: methods, applications, and directions. IEEE Access **12**, 3768–3789 (2024). https://ieeexplore.ieee.org/document/10379639. Accessed 26 June 2024
96. Yang, K., Fan, T., Chen, T., Shi, Y, Yang, Q. (2019) A quasi-Newton method based vertical federated learning framework for logistic regression. arXiv:1912.00513. Accessed 18 June 2024
97. Yang, L., Chai, D., Zhang, J., Jin, Y., Wang, L., Liu, H., Tian, H., Xu, Q., Chen, K.: A survey on vertical federated learning: from a layered perspective (2023). arXiv:2304.01829. Accessed 8 April 2024
98. Yang, Q., Liu, Y., Chen, T., Tong, Y.: Federated machine learning: concept and applications. ACM Trans. Intell. Syst. Technol. **10**(2), 1–19 (2019)
99. Yang, Q., Liu, Y., Cheng, Y., Kang, Y., Chen, T., Yu, H.: Horizontal federated learning. In: Federated Learning. Synthesis Lectures on Artificial Intelligence and Machine Learning. Springer, Cham, Berlin, Germany (2020). https://link.springer.com/chapter/10.1007/978-3-031-01585-4_4. Accessed 8 April 2024
100. Yang, Q., Zhang, Y., Dai, W., Pan, S.J.: Transfer Learning. Cambridge University Press, Cambridge, UK (2020)
101. Yang, S., Ren, B., Zhou, X., Liu, L.: Parallel distributed logistic regression for vertical federated learning without third-party coordinator (2019). arXiv:1911.09824. Accessed 8 April 2024
102. Yin, X., Zhu, Y., Hu, J.: A comprehensive survey of privacy-preserving federated learning: a taxonomy, review, and future directions. ACM Comput. Surv. **54**(6), 1–36 (2021)
103. Yosinski, J., Clune, J., Bengio, Y., Lipson, H.: How transferable are features in deep neural networks? Adv. Neural Inf. Proc. Syst. **27**, 3320–3328 (2014). arXiv:1411.1792. Accessed 26 June 2024
104. Yu, F., Xiu, X., Li, Y.: A survey on deep transfer learning and beyond. Mathematics **10**(19), 3619 (2022). www.mdpi.com/2227-7390/10/19/3619. Accessed 11 March 2024
105. Yuan, L., Wang, Z., Sun, L., S. Yu, P., Brinton, C.G.: Decentralized federated learning: a survey and perspective. IEEE Internet Things J. (2024). https://ieeexplore.ieee.org/document/10542323. Accessed 11 Aug 2024
106. Zantedeschi, V., Bellet, A., Tommasi, M.: Fully decentralized joint learning of personalized models and collaboration graphs. In: Proceeding of International Conference on Artificial Intelligence and Statistics, pp. 864–874 (2020). arXiv:1901.08460. Accessed 8 June 2024

107. Zhang, C., Li, S., Xia, J., Wang, W.: BatchCrypt: efficient homomorphic encryption for cross-silo federated learning. In: Proceedings of the USENIX annual technical conference (2020). www.usenix.org/system/files/atc20-zhang-chengliang.pdf. Accessed 15 April 2024
108. Zhang, C., Xie, Y., Bai, H., Yu, B., Li, W., Gao, Y.: A survey on federated learning. Knowl.-Based Sys. **216**, 106775 (2021). https://doi.org/10.1016/j.knosys.2021.106775. Accessed 19 March 2024
109. Zhao, D., Yao, M., Wang, W., He, H., Jin, X.: Ntp-vfl-a new scheme for non-3rd party vertical federated learning. In: Proceeding of International Conference on Machine Learning and Computing, pp. 134–139 (2022). https://dl.acm.org/doi/10.1145/3529836.3529841. Accessed 11 April 2024
110. Zhou, J., Zhang, S., Lu, Q., Dai, W., Chen, M., Liu, X., Pirttikangas, S., Shi, S., Zhang, W., Herrera-Viedma, E.: A survey on federated learning and its applications for accelerating industrial internet of things (2021). arXiv:2104.10501. Accessed 8 March 2024
111. Zhou, Z., Chu, L., Liu, C., Wang, L., Pei, J., Zhang, Y.: Towards fair federated learning. In: Proceedings of ACM SIGKDD Conference on Knowledge Discovery and Data Mining, Tutorial. Slides (2021). www.cas.mcmaster.ca/~chul9/Contents/KDD_2021_Tutorial.html. Accessed 8 April 2024
112. Zhuang, F., Qi, Z., Duan, K., Xi, D., Zhu, Y., Zhu, H., Xiong, H., He, Q.: A comprehensive survey on transfer learning. Proc. IEEE **109**(1), 43–76 (2021). arXiv:1911.02685. Accessed 11 March 2024

Chapter 5
Data Leakage and Data Poisoning

Abstract This chapter discusses two attacks that target weaknesses of federated learning systems: (1) *data leakage*, inferring raw data used to train an AI model by unauthorized parties and (2) *data poisoning*, a cyberattack that compromises data used to train an AI model to manipulate its output. A major source of data leakage is *gradient sharing* during training of federated learning systems. A dishonest party with access to local gradients from participants during AI model training can carry out a *gradient inversion attack* to reconstruct the raw data of individual participants. As data leakage from gradient sharing came to be recognized as a problem with no easy solution, the practice has been curtailed in commercial deployments. Another threat to federated learning systems is *data poisoning*. Modifying an AI model during training using *poison data* to generate incorrect predictions can be harmful and is particularly problematic when the contaminated global AI model confidently outputs incorrect results that look credible, at a cursory glance. These two attacks demonstrate the difficulty in developing new frameworks for real-world deployments. Since it is beyond the scope of this work to review cryptography, security, and privacy attacks, we refer readers to some survey works.

5.1 Introduction

The previous chapters discussed how federated learning was designed to support collaborative training of machine learning models by multiple parties, while protecting the individual privacy of each party. Each participating party retains full control of its own data, never having to share it with others. Unfortunately, unauthorized parties have succeeded in recovering raw data—or very close approximations for all practical purposes, in the case of images—for some deployments, exemplifying a problem that has come to be known as *data leakage* [6, 23, 44, 48, 49, 54, 65, 67, 83, 86, 95]. A second challenge with federated learning systems is *data poisoning* [1, 16, 18, 35, 46, 66, 72, 78, 82]. Modifying an AI model using poisoned data to generate incorrect predictions can be harmful and is particularly problematic when the contaminated AI model confidently outputs results that look credible, but are incorrect.

M. Kobayashi, *Federated Learning*, ICIAM2023 Springer Series 4,
https://doi.org/10.1007/978-981-96-9223-1_5

These attacks are just two of many types of issues with federated learning that researchers and users face in their daily work. Since it is beyond the scope of this monograph to review the myriad of possible cryptographic, security, and privacy attacks, we refer readers to some survey works. An informative invited talk from ACM KDD 2021 on weaknesses of federated learning systems by Xiao is available on YouTube [83]. Some general surveys on methods for attacking AI systems are [15, 17, 20, 39, 55]. They discuss attacks that are not necessarily limited to federated learning systems. The surveys examine both proactive and reactive defense strategies. As mentioned earlier, in Chaps. 2 and 4, cryptographic defense strategies include homomorphic encryption, secure multiparty computation, and differential privacy. Survey papers on vulnerabilities of federated learning systems include [11, 30, 38, 43, 49, 58]. Some surveys and works on privacy attacks on federated learning systems are [25, 27, 42, 55, 90, 93, 95].

5.2 Data Leakage and Gradient Inversion

The simplest type of data leakage problem concerns multidimensional numerical data. If someone guesses a data point that is an element in a training data set, the confidence level output by the AI model will be 100% or very close, e.g., 99.99998%. The next step would be to identify the owner(s) of the data set(s) containing this data point. It can be easy to infer when only a few institutions are involved in federated learning. For example, with medical data, the number of patients at each hospital is quite small, and the demographics of patients (sex and age) along with disease type might identify a famous celebrity who was spotted by paparazzi entering the doorway of a famous hospital, e.g., Mount Sinai Hospital or Mayo Clinic.

In some instances, raw data can be recovered to various degrees of accuracy using attacks on a model, e.g., extraction of the gender or ethnicity of speakers who trained voice recognition systems [2, 3]. Missing feature inference may be possible when partial feature information is known about data and AI model inversion is used to fill in missing features [23, 24, 87].

A method that has made significant headway in inferring more complex raw data of participants (such as image data) is *gradient inversion*. Gradients of the error (or loss) function for local AI models are computed during AI model training, and the information is shared with the central authority. AI scientists discovered that when the central authority uses this "gradient sharing" information to train or update its global AI model, less computation is needed to create a better global AI model. Unfortunately, if the information on local gradients fall into the wrong hands, it may be possible to reconstruct raw data.

So exactly what are these gradients? Recall from Sect. 1.2 that neural networks are trained on a massive data set, whose correct output is known. The goal is to find weights at each of the nodes of the neural network that will minimize the total or sum of the error in predictions for all of the training data points. The process of finding this magic combination of weights is known as *training an AI model.* This training is a computationally expensive, iterative process that makes AI model construction very expensive.

The training process proceeds as follows. Initially, all of the weights of an AI model are unknown, so they are set to a random, non-zero number (or all are set to 1.0). The total error is computed. To make sure the weights are perturbed in a direction that decreases the error, the discrete, numerical gradient of the error function is computed. Recall from multivariable calculus that the gradient is like the partial derivative, so it equals zero at local and global minima and maxima. The weights are perturbed a little in the direction that decreases the error as fast as possible, and the total error is recomputed. This iterative process continues until one of the following criteria is fulfilled:

- the total error is below a user-specified threshold, i.e., very close to zero;
- the total error changes very little between iterations; and
- the number of iterations exceeds the maximum specified by the user.

More precisely, convergence to the weights that minimize the total error proceeds as follows. For cost function f, and position p_n, we can move in the direction of steepest descent to the next point p_{n+1}, by computing the gradient ∇f of f and setting

$$p_{n+1} = p_n - \eta \nabla f(p_n),$$

where η is a scaling factor for the gradient and controls the step size. If the step size is set to be too small, it will take too much time, with too many steps and computations to converge to the numerical minimum. If the step size is too large, it may bounce off to infinity before it reaches a numerical minimum and diverges. For detailed tutorials on gradient descent and variations that speed up convergence,[1,2] see Chaps. 7 and 8 in [8], [34, 40].

Each participant in a federated learning framework must train its own, local AI model, which is a lot of computational work. The global AI model can be computed by averaging weights from local models. In practice, global AI models must be updated on a regular basis since most systems and input data are in constant flux due to changing conditions (environmental, business, etc.). Sharing gradients of local models during training or updating the global AI model appeared to be a better approach for improving speed and performance than training each local model in isolation. Shortly after gradient sharing was widely adopted for image processing neural networks, algorithms that rely on local gradients kept improving [71]. At that time, researchers never contemplated that it would be possible to recover near-identical

[1] Mehreen Saeed (March 16, 2022) A gentle introduction to gradient descent procedure. Machine Learning Mastery Blog: https://machinelearningmastery.com/a-gentle-introduction-to-gradient-descent-procedure/.

[2] Jason Brownlee (August 12, 2019) Gradient descent for machine learning, Machine Learning Mastery Blog: https://machinelearningmastery.com/gradient-descent-for-machine-learning/.

images to raw image data used for training local neural networks. Unfortunately, their oversight was incorrect and the conclusion was that gradient sharing could lead to severe data leakage.

Recovery of the original data using information in local gradients is known as the *gradient inversion problem*. Initial attempts at gradient inversion succeeded in low-resolution reconstruction of the original image data sets used for training [23, 36, 41, 47, 60]. As research in federated learning accelerated, high-resolution (pixel-level) reconstruction of some image data sets from gradient data became possible for simple images when training with small batch sizes [27, 37].

Recently, neural network inversion methods that can reconstruct training data using gradients with larger batch sizes have been reported [37, 63, 88, 95]. Wang et al. [76] and Zhu et al. [95] report successful recovery of private training data using public gradients for both image and natural language processing. Pan et al. [63] report successful recovery of images for batch sizes up to 32, with theoretical guarantees. Yin at al. [89] report successful recovery of images for batch sizes up to 48, but with no theoretical guarantees. Jin et al. [37] report successful recovery of image data using batch sizes up to 100, with theoretical guarantees. Other works in this area include [9, 11, 22, 88, 93, 94].

Three famous inversion attacks are

- Elastic-net attacks (EAD) generating visually indistinguishable adversarial images using the L_1 norm [14].
- C & W (L_2 norm) generating adversarial examples using the L_2 norm [13].
- Madry (L_1 norm) generating adversarial examples using robust optimization [50].

By no means is this a comprehensive list of gradient-based attacks. Some reviews on the development of gradient inversion methods and attacks are [11, 28, 32, 62, 85].

The success of gradient inversion methods and attacks, particularly for images, has led to research on defending against these attacks. Some recent works in this area include [10, 19, 21, 33, 57, 64, 70, 74, 79–81, 92, 95].

Data leakage from gradient sharing has been problematic for horizontal and vertical federated learning frameworks. Deployments of federated transfer learning have tended to use a massive, open data set or a massive in-house data set owned by a corporation, and only one party with sensitive data is receiving the transferred AI model so data leakage has not been as much of an issue to-date. However, with a recent increase in federated transfer learning, attacks are likely to become an issue.

5.3 Data Poisoning

A second type of common attack on AI and federated learning systems is *data poisoning* [1, 16, 18, 20, 29, 45, 51, 52, 56, 66, 72, 75, 77, 78], in which a malicious actor or party intentionally compromises a data set used for training an AI model by

- injecting fake or misleading data;
- tampering with the fidelity of the training data set; and
- deleting a portion of the data.

Targeted data poisoning attacks aim to output incorrect or misleading results from an AI model for specific situations. Non-targeted attacks aim to degrade the overall performance of an AI model.

Four well-known data poisoning attacks used in the real world by cybercriminals to degrade AI model performance are[3,4]

- *Backdoor poisoning* injects training data to give an attacker access to an AI model via a backdoor and the ability to manipulate the model [61]. Examples include
  - pre-trained AI models that are fine-tuned for a specific task [16] and
  - transfer models [29, 52, 75].
- *Availability attack* disrupts the availability of a system or service [7, 91].
- *Model inversion attack* enables the attacker to re-create raw data.
- *Stealth attack* slowly and covertly injects poison data to degrade an AI model over time to affect output (e.g., inject biases).

Backdoor attacks circumnavigate the authentication procedures of systems to gain unauthorized access through installation of malicious software (e.g., by posing as useful, downloadable freeware), injection of poison data, or exploitation of security weaknesses[5] [61]. Backdoor poisoning is a type of backdoor attack that adds poison data to a training data set to create a point of access, or "backdoor" for an attack. The attacks can be targeted or non-targeted.

In AI model training scenarios, use of transfer learning as a means for backdoor data poisoning attacks is possible, but since the corrupted AI model in transfer learning will be fine-tuned using additional data, the effects of the poisoned data will be diluted and may not be effective as hoped (for example, if the original poison data is small, and very large volumes of data are used for fine-tuning after AI model transfer). Results from a study of the effectiveness of backdoor attacks using transfer learning for image data sets under various conditions are reported in [52].

In availability attacks, the training data is perturbed ever so slightly as to avoid detection, but just enough to modify key properties of the data to change the trained AI model and its output [7]. AI models trained on the poisoned data will have such poor accuracy that they will have little or no practical value. Methods for perturbing input samples for availability attacks have been extensively studied by Yu et al. [91]. They examined mechanisms for rendering data useless for training an AI model to explain how they work.

---

[3] Lenaerts-Bergmans, Bart (March 30, 2024) Crowdstrike website www.crowdstrike.com/cybersecurity-101/cyberattacks/data-poisoning.

[4] Ilmoi, Poisoning attacks on machine learning (July 15, 2019) Towards Data Science blog: https://towardsdatascience.com/poisoning-attacks-on-machine-learning-1ff247c254db.

[5] Lenaerts-Bergmans, Bart (July 17, 2023) Crowdstrike website: https://www.crowdstrike.com/cybersecurity-101/attack-types/backdoor-attack/.

Model inversion attacks use predictions output by an AI model to infer information about the training data and implement a reverse engineering-like approach to reconstruct raw training data.

Data poisoning is fast becoming a serious problem in AI, particularly with the increased use of famous open data sets that are prone to tampering [12]. Tolpegin et al. [73] found that a very small percentage of malicious participants in a federated learning system who engage in data poisoning can cause substantial drops in the global AI model's prediction accuracy and recall. In addition, the attackers can target specific classes for attack. They propose a method to identify these malicious actors. Some recent works on data poisoning attacks against federated learning systems include [4, 5, 31, 52, 69, 84].

Some methods for checking for data poisoning are outlier or anomaly detection, analysis of data distribution, and analysis of the impact of data on an AI model's accuracy. The first two methods may not work if the poison data is within the statistical distribution of the data set as a whole. These methods can also be used for understanding the influence of new data when updating an existing AI model.

Some examples of effective tools to use during AI model updates are

- *Reject on negative impact* (RONI), which tests the impact of each new data point, and rejects data points with a large negative impact [59].
- *Target-aware RONI* (tRONI), a variation of RONI [68].
- *STRIP*, which detects input for trojan attacks during run-time by perturbing the incoming input [26].
- *Human-in-the-loop*, which has professional human security analysts who examine outliers and their effects on the decision boundaries of AI models [53].
- *TRIM*, which estimates regression parameters and uses a trimmed loss function to remove points with large residuals [35].

5.4 Conclusion

The past few decades have seen the acceleration of digitization of data. Institutions seek to leverage the opportunity to gain actionable insights to improve their businesses through increased efficiency, productivity, satisfaction of customers and employees, etc. Until recently, analysis was used to understand past events to plan for the future, but the interest of many decision-makers has shifted to accurate prediction of future events using AI systems. However, for most institutions, the volume of clean, properly formatted data available for training AI systems may be too small, or the processing and storage of massive volumes of data with highly sensitive personal information would be prohibitive.

A variety of methods have been developed to supplement real data, such as data augmentation and synthetic data generation. However, most methods are limited in scope and cannot be applied in some scenarios, such as those involving highly skewed data (e.g., in medical diagnosis and financial fraud). Another possible solution is secure, multiparty computation, in which three or more parties agree to pool together their data for analysis. Unfortunately, multiparty computation cannot be applied to data containing sensitive, personal information due to legal restrictions in many countries, including those in the EU and Japan.

Federated learning is a framework that combines ideas from edge computing and multiparty computation to enable institutions to retain control and possession of their own data by performing all computations with their raw data locally, on their own computer. Only the results from training their own, local AI model (and not the raw data) are shared with other parties for joint training of a global AI model. Federated learning is compliant with privacy regulations in many countries since the raw data with sensitive, personal information never leaves the premises of the owner and is never shared.

Federated learning comes in many flavors. For example:

- Horizontal, cross-device federated learning on mobile devices involving one institution is relatively simple to implement. It is a useful tool for understanding customer behaviors and preferences while respecting the personal privacy of mobile device owners. This type of federated learning relies on edge computing on smart phones, tablets, PCs, and mobile devices of customers.
- Horizontal, cross-silo federated learning is useful when small-to-medium size institutions have the same type of data in a similar format. Although each individual institution has insufficient data for training an AI model on its own, the combined information in their data can be used to build a high-quality AI model.
- Vertical federated learning enables two or more institutions with the same customers, but different stored characteristics jointly construct a broader, holistic view of their customers.
- Federated transfer learning is a possible solution when an institution has insufficient data for training an AI model and there is a reliable AI model trained on a massive data set with somewhat similar data. The reliable AI model is used as a starting point, and it is repurposed after fine-tuning using the small data set. The success of federated transfer learning is difficult to predict. Trial and error and extensive testing are required to evaluate the viability of the technique for each transfer scenario.

From an academic standpoint, federated learning seems like a great solution to address data paucity. However, in practice, federated learning involving several institutions can be difficult to implement since different parties must understand the benefits of joint training of an AI model with a larger volume of data; trust one another; agree to cooperate; ensure the trustworthiness of all associated employees; and ensure the integrity, accuracy, and security of their data and results from local computations (no data or model poisoning).

In conclusion, federated learning is one possible tool to use when there is insufficient data, and when there are multiple institutions with data working in the same domain. To gain wider adoption, federated learning systems must demonstrate that uncovering new, actionable insights exceeds the computational and labor costs associated with its implementation. Furthermore, there must be strong guarantees of data security and fidelity by all parties. Since different institutions and businesses tend to have different data formats, federated transfer learning appears to be the new frontier for research on this topic.

References

1. Ahmed, I.M., Kashmoola, M.Y.: Threats on machine learning technique by data poisoning attack: a survey. In: Abdullah, N., Manickam, S., Anbar, M. (eds.) Advances in Cyber Security. Communications in Computer and Information Science, vol. 1487, pp. 586–600. Springer, Berlin, Germany (2021)
2. Ateniese, G.: Hacking smart machines with smarter ones: how to extract meaningful data from machine learning classifiers. In: Camenisch, J., Manulis, M., Tsudik, G., Wright, R. (eds.) Privacy-Oriented Cryptography (Dagstuhl Seminar 12381). Dagstuhl Report, vol. 2, pp. 165–183 (2012). http://proceedings.mlr.press/v108/bagdasaryan20a.htmldrops.dagstuhl.de/opus/volltexte/2013/3755/pdf/dagrep_v002_i009_p165_s12381.pdf. Accessed 24 June 2024
3. Ateniese, G., Mancini, L. V., Spognardi, A., Villani, A., Vitali, D., Felici, G.: Hacking smart machines with smarter ones: how to extract meaningful data from machine learning classifiers. Int. J. Secur. Netw. **10**(3),137–150 (2015). arXiv:1306.4447. Accessed 24 June 2024
4. Bagdasaryan, E., Veit, A., Hua, Y., Estrin, D., Shmatikov, V.: How to backdoor federated learning. In: Proceedings of International Conference on Artificial Intelligence and Statistics. Published in Proceedings of Machine Learning Research, vol. 108, pp. 2938–2948 (2020). http://proceedings.mlr.press/v108/bagdasaryan20a.html. Accessed 14 Aug 2024
5. Bhagoji, A.N., Chakraborty, S. Mittal, P., Calo, S.: Analyzing federated learning through an adversarial lens. In: Proceedings of International Conference on Machine Learning, PMLR, vol. 97, pp. 634–643 (2019). http://proceedings.mlr.press/v97/bhagoji19a.html. Accessed 4 June 2024
6. Bhowmick, A., John Duchi, J., Freudiger, J., Kapoor, G., Rogers, R.: Protection against reconstruction and its applications in private federated learning (2018). arXiv:1812.00984. Presentation slides http://simons.berkeley.edu/sites/default/files/docs/11421/sdptechnicaltalk2.pdf. Accessed 24 June 2024
7. Biggio, B., Roli, F.: Wild patterns: ten years after the rise of adversarial machine learning. In: Proceedings of ACM KDD Conference on Knowledge Discovery and Data Mining, pp. 2154–2156 (2018). arXiv:1712.03141. Accessed 4 Aug 2024
8. Bishop, C.M., Bishop, H.: Deep Learning: Foundations and Concepts. Springer, Berlin, Germany (2023)
9. Boenisch, F., Dziedzic, A., Schuster, R., Shamsabadi, A. S., Shumailov, I., Papernot, N.: When the curious abandon honesty: federated learning is not private. In: Proceedings of IEEE European Symposium on Security and Privacy (EuroS&P), pp. 175–199 (2023). arXiv:2112.02918. Accessed 24 June 2024
10. Bonawitz, K.A., Ivanov, V., Kreuter, B., Marcedone, A., McMahan, H.B., Patel, S., Ramage, D., Segal, A., Seth, K.: Practical secure aggregation for federated learning on user-held data. In: Proceedings of Neural Information Processing Systems Workshop on Private Multi-Party Machine Learning (2016). arXiv:1611.04482. Accessed 24 June 2024
11. Bouacida, N., Mohapatra, P.: Vulnerabilities in federated learning. IEEE Access **9**, 63229–63249 (2021). http://pmlab.cse.usf.edu/assets/pubs/nader2021vulnerabilities.pdf. Accessed 24 June 2024

12. Carlini, N., Jagielski, M., Choquette-Choo, C.A., Paleka, D., Pearce, W., Anderson, H., Terzis, A., Thomas, K., Tramèr, F.: Poisoning Web-scale training datasets is practical. In: Proceedings of IEEE Symposium on Security and Privacy (SP), p. 179 (2024). arXiv:2302.10149v2. Accessed 30 June 2024
13. Carlini, N., Wagner, D.: Adversarial examples are not easily detected: bypassing ten detection methods. In: Proceedings of ACM Workshop on Artificial Intelligence and Security (AISec '17), pp. 3–14 (2017). http://dl.acm.org/doi/10.1145/3128572.3140444. Accessed 10 Aug 2024
14. Chen, P.Y., Sharma, Y., Zhang, H., Yi, J., Hsieh, C.-J.: EAD: elastic-net attacks to deep neural networks via adversarial examples. In: Proceedings of Association for the Advancement of Artificial Intelligence (AAAI) Conference on Artificial Intelligence, pp. 10–17 (2018). http://cdn.aaai.org/ojs/11302/11302-13-14830-1-2-20201228.pdf. Accessed 8 Aug. 2024
15. Chen, Q., Tao, Y.: An investigation of recent backdoor attacks and defenses in federated learning. In: Proceedings of International Conference on Fog and Mobile Edge Computing (FMEC), pp. 262–269 (2023)
16. Chen, X., Liu, C., Li, B., Lu, K., Song, D.: Targeted backdoor attacks on deep learning systems using data poisoning (2017). arXiv:1712.05526. Accessed 7 Aug. 2024
17. Chen, Y., Gui, Y., Lin, H., Gan, W., Wu, Y.: Federated learning attacks and defenses: a survey. In: Proceedings of IEEE International Conference on Big Data, pp. 4256–4265 (2022)
18. Cinà, A.E., Grosse, K., Demontis, A., Vascon, S., Zellinger, W., Moser, B.A., Oprea, A., Biggio, B., Pelillo, M., Roli, F.: Wild patterns reloaded: a survey of machine learning security against training data poisoning. ACM Comput. Surv. **55**(13s), 1–39 (2023). http://dl.acm.org/doi/10.1145/3585385. Accessed 24 June 2024
19. Du, J., Hu, J., Wang, Z., Sun, P., Gong, N. Z., Ren, K.: SoK: Gradient Leakage in Federated Learning (2024). arXiv:2404.05403. Accessed 24 June 2024
20. Fan, J., Yan, Q., Li, M., Qu, G., Xiao, Y.: A survey on data poisoning attacks and defenses. In: Proceedings of IEEE International Conference on Data Science in Cyberspace (DSC) (2022). http://ieeexplore.ieee.org/document/9900151. Accessed 24 June 2024
21. Fan, M., Liu, Y., Chen, C., Wang, C., Qiu, M., Zhou, W.: Guardian: guarding against gradient leakage with provable defense for federated learning. In: Proceedings of ACM International Conference on Web Search and Data Mining (WSDM), pp. 190–198 (2024)
22. Fang, L., Wang, L., Li, H.: Iterative and mixed-spaces image gradient inversion attack in federated learning. Cybersecurity **7**, 35 (2024). http://doi.org/10.1186/s42400-024-00227-7. Accessed 24 June 2024
23. Fredrikson, M., Jha, S., Ristenpart, T.: Model inversion attacks that exploit confidence information and basic countermeasures. In: Proceedings of ACM SIGSAC Conference on Computer and Communications Security (CCS) (2015). http://dl.acm.org/doi/10.1145/2810103.2813677. Accessed 4 June 2024
24. Fredrikson, M., Lantz, E., Jha, S., Lin, S., Page, D., Ristenpart, T.: Privacy in pharmacogenetics: an end-to-end case study of personalized warfarin dosing. In: Proceedings of USENIX Conference on Security Symposium, pp. 17–32 (2014)
25. Fu, C., Zhang, X., Ji, S., Chen, J., Wu, J., Guo, S., Zhou, J., Liu, A.X., Wang, T.: Label inference attacks against vertical federated learning, pp. 1–15. In: Proceedings of USENIX Security Symposium, Boston, MA, USA (2022)
26. Gao, Y., Xu, C., Wang, D., Chen, S., Ranasinghe, D.C., Nepal, S.: STRIP: a defence against trojan attacks on deep neural networks (2020). arXiv:1902.06531. Accessed 5 Aug 2024
27. Geiping, J., Bauermeister, H., Droge, H., Moeller, M.: Inverting gradients – How easy is it to break privacy in federated learning? Proc. Adv. Neural Inf. Proc. Syst. **33**, 16937–16947 (2020). http://proceedings.neurips.cc/paper_files/paper/2021/hash/c4ede56bbd98819ae6112b20ac6bf145-Paper.pdf. Accessed 4 June 2024
28. Gong, H., Jiang, L., Liu, X., Wang, Y., Gastro, O., Wang, L., Zhang, K., Guo, Z.: Gradient leakage attacks in federated learning. Artif. Intell. Rev. **56**, 1337–1374 (2023)
29. Gu, T., Liu, K., Dolan-Gavitt, B., Garg, S.: BadNets: evaluating backdooring attacks on deep neural networks. IEEE Access (2019). http://ieeexplore.ieee.org/stamp/stamp.jsp?arnumber=8685687. Accessed 5 Aug 2024

30. Hallaji, E., Razavi-Far, R., Saif, M., Wang, B., Yang, Q.: Decentralized federated learning: a survey on security and privacy. IEEE Trans. Big Data **10**(2), 194–213 (2024). arXiv:2401.17319. Accessed 5 Aug 2024. https://openaccess.thecvf.com/content_CVPR_2020/papers/Haroush_The_Knowledge_Within_Methods_for_Data-Free_Model_Compression_CVPR_2020_paper.pdf. Accessed 4 June 2024
31. Hitaj, B., Ateniese, G., Perez-Cruz, F.: Deep models under the GAN: information leakage from collaborative deep learning. In: Proceedings of ACM Conference on Computer and Communications Security (CCS) (2017). arXiv:1702.07464. Accessed 28 June 2024
32. Huang, Y., Gupta, S., Song, Z., Li, K., Arora, S.: Evaluating gradient inversion attacks and defenses in federated learning. In: Proceedings of Neural Information Processing Systems Conference (2021). http://openreview.net/forum?id=0CDKgyYaxC8. Accessed 24 June 2024
33. Huang, Y., Song, Z., Li, K., Arora, S.: Instahide: Instance-hiding schemes for private distributed learning. In: Proceedings of International Conference on Machine Learning (2020). arXiv:2010.02772. Accessed 24 June 2024
34. IBM website, What is gradient descent? www.ibm.com/topics/gradient-descent. Accessed 2 Aug 2024
35. Jagielski, M., Oprea, A., Biggio, B., Liu, C., Nita-Rotaru, C., Li, B.: Manipulating machine learning: poisoning attacks and countermeasures for regression learning (2021). arXiv:1804.00308. Accessed 10 Aug 2024
36. Jensen, C.A., Reed, R.D., Marks, R.J., El-Sharkawi, M.A., Jung, J.-B., Miyamoto, R.T., Anderson, G.M., Eggen, C.J.: Inversion of feedforward neural networks: algorithms and applications. Proc. IEEE **87**(9), 1536–1549 (1999)
37. Jin, X., Chen, P.-Y., Hsu, C.-Y., Yu, C.-M., Chen, T.: CAFE: catastrophic data leakage in vertical federated learning. Adv. Neural Inf. Proc. Syst. **34** (2021). http://proceedings.neurips.cc/paper_files/paper/2016/file/08040837089cdf46631a10aca5258e16-Paper.pdf. Accessed 4 June 2024
38. Kairouz, P., McMahan, B., Avent, B., Bellet, A., Bennie, M. et al.: Advances and open problems in federated learning. Found. Trends Mach. Learn. **14**(1–2), 1–210 (2021). arXiv:1912.04977. Accessed 10 April 2024
39. Kumar, K.N., Mohan, C.K., Cenkeramaddi, L.R.: The impact of adversarial attacks on federated learning: a survey. IEEE Trans. Pattern Anal. Mach. Intell. **46**(5), 2672–2691 (2024)
40. Kwiatkowski, R.: Gradient Descent Algorithm, a deep dive. Towards Data Science (2021). http://towardsdatascience.com/gradient-descent-algorithm-a-deep-dive-cf04e8115f21. Accessed 5 Aug. 2024
41. Lee, S., Kil, R.M.: Inverse mapping of continuous functions using local and global information. IEEE Trans. Neural Netw. **5**(3), 409–423 (1994)
42. Li, Z., Sharma, V., Mohanty, S.P.: Preserving data privacy via federated learning: challenges and solutions. IEEE Consumer Electron. Mag. **9**(3), 8–16 (2020). www.smohanty.org/Publications_Journals/2020/Mohanty_IEEE-MCE_2020-May_Federated-Learning-Privacy.pdf. Accessed 9 Aug 2024
43. Lim, W.Y.B., Luong, N.C., Hoang, D.T., Jiao, Y., Liang, Y.-C., Yang, Q., Niyato, D., Miao, C.: Federated learning in mobile edge networks: a comprehensive survey. IEEE Commun. Surv. Tutor. **22**(3), 2031–2063 (2020)
44. Linden, A., Kindermann, J.: Inversion of multilayer nets. In: Proceedings of International Joint Conference on Neural Networks, vol. 2, pp. 425–430 (1989)
45. Liu, J., Huang, J., Zhou, Y., Li, X., Ji, S., Xiong, H., Dou, D.: From distributed machine learning to federated learning: a survey. Knowl. Inf. Syst. **64**(4), 885–917 (2022). arXiv:2104.14362. Accessed 9 Aug 2024
46. Liu, Y., Mondal, A., Chakraborty, A., Zuzak, M., Jacobsen, N., Xing, D., Srivastava, A.: A survey on neural Trojans. In: Proceedings of International Symposium on Quality Electronic Design (ISQED), pp. 33–39 (2020)
47. Lu, B.-L., Kita, H., Nishikawa, Y.: Inverting feedforward neural networks using linear and nonlinear programming. IEEE Trans. Neural Netw. **10**(6), 1271–1290 (1999)

48. Lyu, L., Yu, H., Ma, X., Chen, C., Sun, L., Zhao, J., Yang, Q., Yu, P.: Privacy and robustness in federated learning: attacks and defenses. IEEE Trans. Neural Netw. Learn. Syst. (2022). arXiv:2012.06337. Accessed 25 June 2024
49. Lyu, L., Yu, H., Yang, Q.: Threats to federated learning: a survey (2020). arXiv:2003.02133. Accessed 24 June 2024
50. Madry, A., Makelov, A., Schmidt, L., Tsipras, D., Vladu, A.: Towards deep learning models resistant to adversarial attacks. In: Proceedings of International Conference on Learning Representations (ICLR), Poster (2018). arXiv:1706.06083. Accessed 4 Aug 2024
51. Mammen, P.M.: Federated learning: opportunities and challenges (2021). arXiv:2101.05428. Accessed 7 Aug 2024
52. Matsuo, Y., Takemoto, K.: Backdoor attacks on deep neural networks via transfer learning from natural images. Appl. Sci. **12**(24), 12564 (2022). http://doi.org/10.3390/app122412564. Accessed 4 Aug 2024
53. Mei, S., Zhu, X.: Using machine teaching to identify optimal training-set attacks on machine learners. In: Proceedings of AAAI Conference on Artificial Intelligence, vol. 29, no. 1 (2015). http://ojs.aaai.org/index.php/AAAI/article/view/9569. Accessed 4 Aug 2024
54. Melis, L., Song, C., Cristofaro, E. D., Shmatikov, V. (2019) Exploiting unintended feature leakage in collaborative learning. In: Proceedings of IEEE Symposium on Security and Privacy, pp. 691–706. arXiv:1805.04049. Accessed 4 June 2024
55. Moore, E., Imteaj, A., Rezapour, S., AMini, M. H.: A survey on secure and private federated learning using blockchain: theory and application in resource-constrained computing. IEEE Int. Things J. **10**(24), 21942–21958 (2023)
56. Mothukuri, V., Parizi, R.M., Pouriyeh, S., Huang, Y., Dehghantanha, A., Srivastava, G.: A survey on security and privacy of federated learning. Future Gener. Compu. Syst. **115**, 619–640 (2021). www.sciencedirect.com/science/article/abs/pii/S0167739X20329848. Accessed 4 Aug. 2024
57. Nair, A.K., Raj, E.D., Sahoo, J.: A robust analysis of adversarial attacks on federated learning environments. Comput. Stand. Interf. **86**, 103723 (2023)
58. Narula, M., Meena, J., Vishwakarma, D.K.: A comprehensive review on federated learning for data-sensitive application: open issues & challenges. Eng. Appl. Artif. Intell. **133**, Part B, 108128 (2024)
59. Nelson, B., Barreno, M., Chi, F.J., Joseph, A.D., Rubinstein, B.I.P., Saini, U., Sutton, C., Tygar, J.D., Xia, K.: Misleading learners: co-opting your spam filter. In: Tsai, J., Yu, P. (eds.) Machine Learning in Cyber Trust: Security, Privacy, Reliability. Springer, Berlin, Germany, pp. 17–51 (2009). http://people.eecs.berkeley.edu/~tygar/papers/SML/misleading.learners.pdf. Accessed 4 Aug. 2024
60. Nguyen, A. Dosovitskiy, A., Yosinski, J., Brox, T., Clune, J.: Synthesizing the preferred inputs for neurons in neural networks via deep generator networks. In: Proceedings of Neural Information Processing Systems (2016). http://proceedings.neurips.cc/paper_files/paper/2016/file/5d79099fcdf499f12b79770834c0164a-Paper.pdf. Accessed 10 June 2024
61. Nguyen, T.D., Nguyen, T., Nguyen, P.L., Pham, H.H., Doan, K.D., Wong, K.-S.: Backdoor attacks and defenses in federated learning: survey, challenges and future research directions. Eng. Appl. Artif. Intell. **127**, Part A, 107166 (2024). arXiv:2303.02213. Accessed 10 Aug 2024
62. Ovi, P.R., Gangopadhyay, A.: A comprehensive study of gradient inversion attacks in federated learning and baseline defense strategies. In: Proceedings of IEEE Conference on Information Sciences and Systems (CISS), pp. 1–6 (2023)
63. Pan, X., Zhang, M., Yan, Y., Zhu, J., Yang, M.: Theory-oriented deep leakage from gradients via linear equation solver (2020). arXiv:2010.13356. Accessed 10 June 2024
64. Phong, L.T., Aono, Y., Hayashi, T., Wang, L., Moriai, S.: Privacy-preserving deep learning via additively homomorphic encryption. IEEE Trans. Inf. Forens. Secur. **13**(5), 1333–1345 (2018)
65. Rafi, T.H., Noor, F.A., Hussain, T., Chae, D.-K.: Fairness and privacy-preserving in federated learning: a survey. Inf. Fus. 105 (2024)
66. Ramirez, M.A., Kim, S.-K., Hamadi, H.A., Damiani, E., Byon, Y.-J., Kim, T.-Y., Cho, C.-S., Yuen, C.Y.: Poisoning attacks and defenses on artificial intelligence: a survey (2022). arXiv:2202.10276. Accessed 30 June 2024

67. Shokri, R., Stronati, M., Song, C., Shmatikov, V.: Membership inference attacks against machine learning models. In: Proceedings of IEEE Symposium on Security and Privacy, pp. 3–18 (2017). arXiv:1610.05820. Accessed 4 June 2024
68. Suciu, O., Marginean, R., Kaya, Y., Daumé, H., Dumitras, T.: When does machine Learning FAIL? Generalized transferability for evasion and poisoning attacks. In: Proceedings of USENIX Conference on Security Symposium, pp. 1299–1316 (2018). www.usenix.org/system/files/conference/usenixsecurity18/sec18-suciu.pdf. www.usenix.org/conference/usenixsecurity18/presentation/suciu. Accessed 4 Aug. 2024
69. Sun, G., Cong, Y., Dong, J., Wang, Q., Liu, J.: Data poisoning attacks on federated machine learning. IEEE Int. Things J. **9**(13), 11365–11375 (2022). arXiv:2004.10020. Accessed 24 June 2024
70. Sun, J., Li, A., Wang, B., Yang, H., Li, H., Chen, Y., Soteria: Provable defense against privacy leakage in federated learning from representation perspective. In: Proceedings of IEEE/CVF Conference on Computer Vision and Pattern Recognition (CVPR), pp. 9311–9319 (2021). http://openaccess.thecvf.com/content/CVPR2021/papers/Sun_Soteria_Provable_Defense_Against_Privacy_Leakage_in_Federated_Learning_From_CVPR_2021_paper.pdf. Accessed 4 Aug 2024
71. Tenison, I., Sreeramadas, S.A., Mugunthan, V., Oyallon, E., Rish, I., Belilovsky, E.: Gradient masked averaging for federated learning. Trans. Mach. Learn. Res. (2023). http://openreview.net/pdf?id=REAyrhRYAo. Accessed 24 June 2024
72. Tian, Z., Cui, L., Liang, J., Yu, S.: A comprehensive survey on poisoning attacks and countermeasures in machine learning. ACM Comput. Surv. **55**(8) 1–35 (2022). http://dl.acm.org/doi/10.1145/3551636. Accessed 24 June 2024
73. Tolpegin, V., Truex, S., Gursoy, M.E., Liu, L.: Data poisoning attacks against federated learning systems. In: Proceedings of European Symposium on Research in Computer Security (ESORICS), Part 1, pp. 480–501 (2020). arXiv:2007.08432. Accessed 30 June 2024
74. Wang, J., Guo, S., Xie, X., Qi., H.: Protect privacy from gradient leakage attack in federated learning. In: Proceedings of IEEE Conference on Computer Communications (INFOCOM), pp. 580–589 (2022)
75. Wang, S., Nepal, S., Rudolph, C., Grobler, M., Chen, S., Chen, T.: Backdoor attacks against transfer learning with pre-trained deep learning models. IEEE Trans. Serv. Comput. **15**(3), 1526–1539. arxiv.org/pdf/2001.03274. Accessed 7 Aug 2024
76. Wang, Y., Deng, J., Guo, D., Wang, C., Meng, X., Liu, H. Ding, C., Rajasekaran, S.: SAPAG: a self-adaptive privacy attack from gradients (2020). arXiv:2009.06228. Accessed 10 June 2024
77. Wang, Z., Kang, Q., Zhang, X., Hu, Q.: Defense strategies toward model poisoning attacks in federated learning: a survey. In: Proceedings of IEEE Wireless Communications and Networking Conference (WCNC), pp. 548–553 (2022). arXiv:2202.06414. Accessed 7 Aug 2024
78. Wang, Z., Ma, J., Wang, X., Hu, J., Qin, Z., Ren, K.: Threats to training: a survey of poisoning attacks and defenses on machine learning systems. ACM Comput. Surv. **55**(7), 1–36 (2022). http://dl.acm.org/doi/10.1145/3538707. Accessed 30 June 2024
79. Wei, W., Liu, L., Loper, M., Chow, K.-H., Gursoy, M.E., Truex, S., Wu, Y.: A framework for evaluating gradient leakage attacks in federated learning. In: Proceedings of European Symposium on Research in Computer Security (ESORICS), pp. 545–566 (2020). arXiv:2004.10397. Accessed 10 June 2024
80. Wei, W., Liu, L., Wu, Y., Su, G., Iyengar, A.: Gradient-leakage resilient federated learning. In: Proceedings of IEEE International Conference on Distributed Computing Systems (ICDCS), pp. 797–807 (2021). www.researchgate.net/publication/353053763_Gradient-Leakage_Resilient_Federated_Learning. Accessed 10 Aug 2024
81. Wu, J., Hayat, M., Zhou, M., Harandi, M.: Concealing sensitive samples against gradient leakage in federated learning. In: Proceedings of AAAI Conference on Artificial Intelligence, vol. 38, no. 19, pp. 21717–21725 (2024). http://ojs.aaai.org/index.php/AAAI/article/view/30171. Accessed 10 Aug 2024

82. Xia, G., Chen, J., Yu, C., Ma, J.: Poisoning attacks in federated learning: a survey. IEEE Access **11**, 10708–10722 (2023)
83. Xiao, X.: Preserving Data Privacy in Federated Learning, Invited Speaker, ACM SIGKDD International Conference on Knowledge Discovery and Data Mining (2021). www.youtube.com/watch?v=9y6g4dE9kzA. Accessed 8 April 2024
84. Xie, C., Huang, K., Chen, P., Li, B.: DBA (2020) Distributed backdoor attacks against federated learning. In: Proceedings of International Conference on Learning Representations (2020). http://iclr.cc/virtual_2020/poster_rkgyS0VFvr.html. Accessed 10 June 2024
85. Yang, H., Ge, M., Xue, D., Xiang, K., Li, H., Lu, R.: Gradient leakage attacks in federated learning: research frontiers, taxonomy, and future directions. IEEE Netw. **38**(2), 247–254 (2024)
86. Yang, Z., Chang, E.-C., Liang, Z.: Neural network inversion in adversarial setting via background knowledge alignment. In: Proceedings of ACM SIGSAC Conference on Computer and Communications Security, pp. 225–240 (2019). Earlier version arXiv:1902.08552. Accessed 4 June 2024
87. Yeom, S., Giacomelli, I., Fredrikson, M., Jha, S.: Privacy risk in machine learning: analyzing the connection to overfitting. In: Proceedings of IEEE Computer Security Foundations Symposium (CSF), pp. 268–282 (2018). arXiv:1709.01604. Accessed 10 June 2024
88. Yin, H., Mallya, A., Vahdat, A., Alvarez, J.M., Kautz, J., Molchanov, P.: See through gradients: image batch recovery via GradInversion. In: Proceedings of IEEE Conference on Computer Vision and Pattern Recognition, pp. 16337–16346 (2021). arXiv:2104.07586. Accessed 4 June 2024
89. Yin, H. Molchanov, P., Alvarez, J.M., Li, Z., Mallya, A. Hoiem, D., Jha, N.K., Kautz, J.: Data-free knowledge transfer via DeepInversion (2019). arXiv:1912.08795. Accessed 24 June 2024
90. Ying, Z., Zhang, Y., Liu, X.: Privacy-preserving in defending against membership inference attacks. In: Proceedings of Workshop on Privacy-Preserving Machine Learning in Practice, pp. 61–63 (2020). http://dl.acm.org/doi/10.1145/3411501.3419428. Accessed 9 Aug 2024
91. Yu, D., Zhang, H., Chen, W., Yin, J., Liu, T.-Y.: Availability attacks create shortcuts. In: Proceedings of ACM KDD Conference on Knowledge Discovery and Data Mining, pp. 2367–2376 (2022). arXiv:2111.00898. Accessed 4 Aug 2024
92. Zhang, H., Cisse, M., Dauphin, Y.N., Lopez-Paz, D.: Mixup: beyond empirical risk minimization. In: Proceedings of International Conference on Learning Representations (ICLR) (2018). arXiv:1710.09412. Accessed 10 June 2024
93. Zhao, B., Mopuri, K.R., Bilen, H.: iDLG: Improved deep leakage from gradients (2020). arXiv:2001.02610. Accessed 10 June 2024
94. Zhu, J., Blaschko, M.B.: R-GAP: recursive gradient attack on privacy. In: Proceedings of International Conference on Learning Representations (2021). http://openreview.net/pdf/943e8621d1ff1fb8b9daa8cc68876c327966099c.pdf. Accessed 10 July 2024
95. Zhu, L., Liu, Z., Han, S.: Deep leakage from gradients. In: Wallach, H., Larochelle, H., Beygelzimer, A., Alché-Buc, F. D., Fox, E., Garnett, R. (eds.) Proceeding of Advances in Neural Information Processing Systems, vol. 32, pp. 14774–14784 (2019). http://papers.nips.cc/paper_files/paper/2019/hash/60a6c4002cc7b29142def8871531281a-Abstract.html. Accessed 10 July 2024

Index

M. Kobayashi, *Federated Learning*, ICIAM2023 Springer Series 4,
https://doi.org/10.1007/978-981-96-9223-1

MIX
Papier aus verantwortungsvollen Quellen
Paper from responsible sources
FSC® C105338

If you have any concerns about our products,
you can contact us on
ProductSafety@springernature.com

In case Publisher is established outside the EU,
the EU authorized representative is:
Springer Nature Customer Service Center GmbH
Europaplatz 3, 69115 Heidelberg, Germany

Printed by Libri Plureos GmbH
in Hamburg, Germany